MINISTRY OF
AGRICULTURE, FISHERIES AND FOOD

# Sugar Beet Pests

by

## F. G. W. JONES, M.A., Sc.D.

Deputy Director,
Rothamsted Experimental Station,
and
Head of Nematology Department

and

## R. A. DUNNING, M.Sc., Ph.D.

Entomologist,
Broom's Barn Experimental Station

*Bulletin 162*

*LONDON*
HER MAJESTY'S STATIONERY OFFICE
1972

SBN 11 241462 1

# Acknowledgments

WE gratefully acknowledge the help given by our colleagues at Rothamsted, Broom's Barn and in A.D.A.S., especially with the section on Docking disorder. Dr. A. E. Boyd kindly advised on strangles, Mr. W. E. Bray on herbicide injury, and the Infestation Control Laboratory on birds and mammals. Drawings and photographs are acknowledged individually below.

Special thanks are due to the Editor, Miss M. Gratwick, who, by her patience and attention to detail, has been able to produce a uniform account from those of the joint authors.

F. G. W. JONES

R. A. DUNNING

# Acknowledgments for Illustrations

*Figures*

Drawings were kindly supplied by:

| | |
|---|---|
| Marie D. Winder . . . | Figures 4(a), (b), (c) and (d) |
| G. D. Heathcote . . . . | Figures 10(a) and (b) |

*Plates*

Photographs were kindly supplied by:

| | | |
|---|---|---|
| Broom's Barn Experimental Station | Plate III | bottom left |
| | VIII | bottom left |
| | X | top, and bottom left and right |
| | XI | middle |
| | XII | middle |
| W. E. Dant . . . . | Plate IV | top left and right, and bottom right |
| | VII | top left |
| | VIII | top right |
| C. C. Doncaster . . . . | Plate VIII | bottom right |
| | IX | top left and bottom left |
| Institutes für Zuckerrübenforschung, Göttingen | Plate III | bottom right |
| Rothamsted Experimental Station | Plate VII | top right, and bottom left and right |
| | VIII | top left |
| | IX | top right |
| G. H. Winder . . . . | Plate III | top right |
| | IV | bottom left |

iii

# Foreword

THE SUGAR BEET crop occupies a prominent place in the agricultural economy of the country and the search for improvements in all aspects of its cultivation commands the continuing attention of research workers and advisers alike.

The recent increase in the practice of 'drilling to a stand' of necessity requires the obvious essential need to preserve the resulting seedlings from the many hazards surrounding them. Of these the numerous pests of the crop are of great importance and to avoid the losses they may cause from germination to harvesting of the roots demands their quick recognition and the timely application of the appropriate control measures.

This Bulletin, which aims at providing such information, has been brought up to date by Dr. F. G. W. Jones of Rothamsted Experimental Station and Dr. R. A. Dunning of the associated Broom's Barn Experimental Station and replaces the edition of 1969 for which they were responsible. Once again thanks are due to those colleagues whose work has provided new information and particularly to the Sugar Beet Research and Education Committee whose interest and encouragement both financial and otherwise continue to be invaluable in increasing the productivity of this crop.

M. COHEN,
*Director,*
*Plant Pathology Laboratory,*
*Harpenden, Herts.*

*Ministry of Agriculture,*
*Fisheries and Food*
December, 1971

# Contents

Asterisks denote the relative importance of each pest, importance increasing with the number of asterisks.

Asterisks denote the relative importance of each pest, importance increasing with the number of asterisks.

# Introduction

THIS Bulletin is concerned primarily with pests of sugar beet, but also deals with mangel, fodder beet, red beet, spinach beet, chard and spinach. All these plants, except the last, are varieties derived from the same wild species, *Beta vulgaris* L., of which the sub-species *maritima* occurs on European coasts just beyond the reach of high tides.

Reference to early works on crop pests shows that many of the present insect pests of sugar beet were recognized on mangels a hundred years ago. The extension of the beet acreage since 1925 has not led to the appearance of any important new insect pest of beet, although intensive beet growing in the 1930s led to much damage by pygmy beetle. Even beet cyst eelworm, which was not observed on sugar beet in Great Britain before 1934, is known to have occurred earlier on mangels. The area found infested with this pest has steadily increased, especially in the Fens. Since 1956 losses from wireworms and rabbits have decreased, but those from various birds have increased. In recent years stunting of beet by free-living eelworms on light land (Docking disorder) seems to have become more widespread and more severe in some seasons; this may be due to intensive cereal growing rather than to 40 years of beet growing.

## CRITICAL STAGES FOR PEST ATTACK

Sugar beet, like some other root crops, passes through a delicate seedling stage during which it is very susceptible to injury. It lacks the early vigour given to the potato crop by reserves of food in the seed tubers, and cannot compensate fully for loss of plants as can crops with many more plants per acre, especially cereals which, after establishment, produce extra stems by tillering. Hence, it is not surprising that there are many pests that can greatly harm the young sugar beet crop, particularly between germination and the four rough-leaf stage; pests that damage the growing point or the tap root are especially serious. At the two to four leaf stage the skin of the stem just below soil level splits, exposing soft inner tissue; damage by soil pests such as pygmy beetle and millepedes is often concentrated at this site.

Seed 'drilled-to-a-stand', which inevitably means a full herbicide programme, produces widely spaced seedlings and virtually no weeds. The seedlings rarely seem to grow as vigorously as do more closely spaced seedlings without herbicide, and more pests tend to congregate on them. A full braird, singled carefully to discard damaged or weakly seedlings, produces the best and most vigorous plant population but is more costly to establish.

Once the beet crop is beyond the six to eight rough-leaf stage and is growing well, it is generally safe from the various seedling pests, although most of these can still be found in the crop and are sometimes numerous. The serious pests of the crop after the seedling stage are those that transmit viruses, those that are large and seriously injure or kill plants by their feeding, as do some birds and mammals, and those that can multiply rapidly and cause damage because they are so numerous, such as black bean aphid.

# PREVENTION OF DAMAGE: GOOD HUSBANDRY

ROTATION

The term 'good husbandry' usually includes a rotation of different crops and this is one of the main safeguards against some, but not all, soil-inhabiting pests. Whether it is effective depends on the life history and host crops of the pest; for instance, wireworms, which can seriously harm beet, are favoured by grass and frequent cereal growing, so that growing these crops in previous years must also be taken into account. Against highly mobile pests, e.g., aphids and birds, crop rotation offers no protection, but it is effective against relatively immobile pests with a narrow host range, e.g., beet cyst eelworm, and is partially effective against insects with at least one stage inhabiting the soil, e.g., pygmy beetle.

Several of the country's most serious pests have become so because of too frequent cropping with susceptible crops, e.g., potato cyst eelworm in the silt areas of Lincolnshire and beet cyst eelworm in the black fen areas of the Isle of Ely. For sugar beet and its allies, not more than one crop in every four years is desirable. Growers' contracts with beet sugar factories contain a rotational clause (Clause 13) and in certain areas rotation is enforced by the Beet Eelworm Orders, 1960 and 1962 (see page 78).

SOIL FERTILITY

Before sugar beet is sown the land should be in good heart and in good tilth; this is best achieved by attention during previous cropping. For the beet crop itself, manuring should be adequate and balanced and any deficiencies of minor elements made good; where necessary, magnesium and boron should be applied to the seedbed and manganese sprayed on the plants at an early stage, i.e., in May or June, (boron can also be applied as a foliar spray in June or July). Sugar beet cannot tolerate acid soil so that acidity must be corrected by liming, preferably early to ensure adequate mixing in the soil. The British Sugar Corporation or the Agricultural Development and Advisory Service will advise on these matters.

DRILLING AND SUBSEQUENT CULTIVATIONS

Firm and fine soil around the seed improves germination and minimizes damage by some soil pests. Seed must be drilled deep enough, 'in the moisture', but drilling too deep is a common cause of weak seedlings that succumb readily to slight pest damage. It is not enough to set the drill in the workshop or yard; the actual depth of drilling must be checked in the field. Herbicide sprays must be applied at the correct rate, otherwise there is a risk of retarding seedling growth and thereby increasing the effect of pest attack. Seed rates are discussed below, but they can be increased a little with advantage where such pests as wireworms are expected, especially when sowing is very early. Early sowing increases yields partly because large plants are produced early in the season and these are better able to withstand attack by some pests. When pests are damaging seedlings in a braird that is to be singled, singling is best delayed. Extra nitrogen is sometimes advocated to stimulate growth of plants damaged by pests; this can be justified only where nitrogen deficiency is probable, for example on light sandy soils where heavy rainfall has caused leaching.

It now seems doubtful whether inter-row cultivation is of any benefit to plant growth on most soils. Steerage hoeing too close to the seedlings can be very injurious; the resulting damage and losses are aptly attributed to 'steelworm'. An adequate and uniform seedling population is a prerequisite of a full stand of plants; the crop will not suffer from a few losses provided they are fairly uniformly distributed.

On land deficient in lime, infertile, or in a poor physical state, beet grows slowly and is likely to suffer from pest attack. The correct cure for this is better farming; insecticides can only help to save what will inevitably be a poor crop. Vigorous crops recover remarkably well from pests, especially those that attack seedlings. Adversities such as cold soil or drought, which often enhance pest damage, cannot readily be countered and insecticides may then help the crop through a difficult period.

## PRINCIPLES OF CONTROL OF BEET PESTS

### PLANT POPULATIONS AND PEST NUMBERS

Pest problems of sugar beet or any other crop depend largely on the relation between pest numbers and plant numbers, but are of course often modified by the vigour of plant growth. Control of pests means decreasing the numbers damaging the crop to a point where little harm is done.

Many field trials have shown that the best sugar yield is obtained from a plant population of about 30,000 per acre regularly spaced; in practice the national average is around 25,000 per acre at the present time. The value of populations as great as 50,000 per acre is being examined, since new methods of growing the crop are leading to a more irregularly spaced plant population that may occasionally be as great as this.

Although populations of established plants have not changed over the years, seedling populations continue to decline steadily. In the early days of the industry only natural multigerm seed was used; about 400,000 seedlings per acre grew thickly and irregularly in the rows, and randomly distributed losses from pests and diseases caused little concern. Rubbed multigerm seed was introduced in the early 1950s, and from 1954 to 1971 the acreage of beet sown with precision drills increased from 4 to 99 per cent. Today most crops are sown with pelleted seed, either genetic monogerm or polyploid multigerm. The pelleting makes it easier to drill the seed and also ensures that each seed has its correct dosage of insecticide. When seed is spaced 5 in. apart in rows 20 in. apart, which is an average spacing if avoiding very early drilling and aiming to achieve minimum labour in the crop, 63,000 seeds are sown per acre. Failure to germinate and losses from pests, diseases and physical causes can greatly decrease the number of seedlings that emerge. Pelleted monogerm seed sown at different spacings at widely scattered sites on different soils in 1970 and 1971 gave a seedling establishment range of 24–71 per cent and 32–73 per cent respectively. Seedling counts in 1965 in nine fields with seed spacings of $1\frac{1}{2}$–2 in. (average 1·8 in.) showed that 75,000–219,000 (average 142,000) seedlings emerged. In 1966 somewhat fewer seedlings were obtained at the same seed spacings, whilst in sixteen other fields with seed spacings of 3–6 in. (average 4·8 in.) 32,000–177,000 (average 56,000) seedlings emerged. These results emphasize the very considerable range in seedling emergence that can occur, partly due to pests and diseases but probably mainly due to physical factors affecting germination. During these observations, seedling losses due to all causes were recorded: most losses were caused by physical factors (steerage hoeing, herbicides, etc.) followed by bird, mammal and insect pests. Total losses up to the time of singling varied enormously from field to field, ranging from a few hundred up to as many as 42,000 per acre. Such data perhaps exaggerate the difficulty of deciding in March what seed spacing is necessary to ensure a satisfactory plant population in June; in practice, an increasing proportion of the crop is being sown at wide spacing and a satisfactory plant population achieved with little or no singling. Herbicides are applied almost universally; the combination of pre- and post-emergence sprays, which eliminates hand hoeing for weed control, has hastened the trend to wide seed-spacing.

An alternative to wide seed-spacing is to sow more thickly and to thin mechanically; this allows more latitude for seedling losses but still produces a poorer final plant stand than does good hand-singling. Nevertheless, there are many who feel that 'drilling-to-a-stand' is preferable, at least on soils that give good seedbeds. Some seedling pests are always likely to be present and may tend to congregate on seedlings, especially when weeds are few. More pests per seedling mean more risk of serious damage to or death of the seedling, yet there are few or no seedlings to spare. In addition, virus infections in the seedling stage will tend to be proportionately more numerous.

The approximate number of pests per acre or per plant above which damage is likely to ensue is specified wherever possible in this Bulletin. It serves as a rough practical guide to the need for control measures, but these critical numbers need decreasing appropriately where there are fewer than 30,000 seedlings emerging per acre.

In the early 1950s, field trials were made to assess the effects of defoliation and loss of stand on yield, so as to indicate the best course of action when attacks occur and the amount that could justifiably be spent on direct control measures. At the four and eight rough-leaf stages, plants were defoliated to varying extents and plant losses after singling were simulated by removing plants in a regular manner, e.g., every fourth plant along the row to give 25 per cent decrease. The main effect of these treatments was to modify the size of the plants; there was little effect upon sugar content. Removing half the leaves had little effect on yield (Fig. 1) whether done at the four or eight rough-leaf stage. The worst possible treatment, i.e., complete defoliation leaving only the growing point, decreased yield by an average of 27 per cent; defoliation to this extent is rare in field crops. Removing whole plants was more serious, but a high degree of compensation occurred because the plants were removed individually and not in groups; pest attack is often different, with patches of affected plants, and then compensation is less (Fig. 2). The trials emphasized that re-drilling in May is best avoided, provided that half or more of the intended plant population remains reasonably distributed over the land.

Field trials in 1966–69 tested complete defoliation at monthly intervals during the growing season. Maximum decrease of sugar yield, nearly 40 per cent, followed July or August defoliation; earlier or later removal of all leaves except the growing point had less effect on yield.

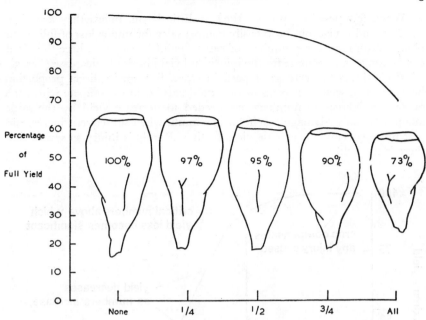

Fig. 1. The effect of defoliation upon yield of roots

Results of six field trials in which the plants were defoliated artificially in the 4- to 8-leaf stage

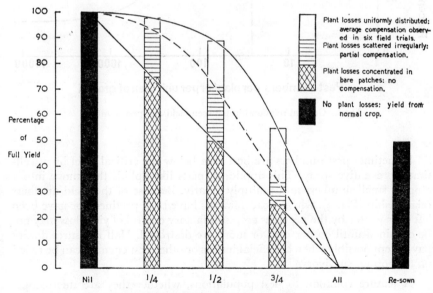

Fig. 2. The effect of plant losses upon yield of roots

*The upper curve* is based on the results of six field trials in which plant losses were simulated by hoeing out different proportions of the full stand shortly after singling. *The lower curve* (a straight line) shows the relation between plant population and yield in the absence of compensation. *The broken line* shows the effect of irregular losses as caused by pests.

When few pests are present they do little harm, unless they are virus carriers, and a vigorous crop easily compensates for minor loss of foliage or plants. With increasing numbers of pests a point is reached where the plant can no longer compensate for the injury and yield begins to decrease (Fig. 3). Yield continues to decrease as pests increase, but not in direct proportion because the insects, eelworms or other animals begin to compete with each other and increasing numbers are needed to decrease yield by the same amount. Finally, damage reaches a maximum when subsequent increase in pest numbers cannot decrease yield further. Rarely is injury so severe that there is no yield.

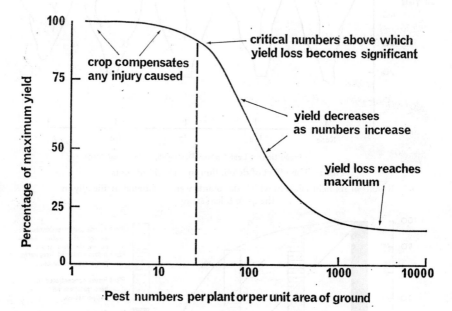

Fig. 3. General relationship between yield and pest numbers

Sometimes pest numbers are best kept below the critical level by a well-timed preventive spray. This applies to pests like aphids that move into a crop in small numbers and multiply greatly. Because of the yield/numbers relationship (Fig. 3), when great numbers have been produced or have been left over winter by the previous crop satisfactory control of yield loss may not be obtained until 90 per cent or more are destroyed. Half measures do not pay except possibly when beneficial insects or other pest enemies can be relied on to mop up survivors.

A feature common to pest populations, whether they are microscopic eelworms, insects, or warm-blooded animals like the rabbit or wood-pigeon, is that they reproduce fastest when populations are smallest. Successful control measures usually bring populations to the state of maximum reproduction so that, when control is relaxed or its effects wear off, the former population is soon restored or even exceeded. Pests are so resilient that their extermination is nearly impossible: control has to be repeated indefinitely.

## NATURAL CONTROL

Natural control of pests is that measure of population limitation exerted by climate, competition and naturally occurring enemies (see page 105). Unfortunately, although it limits crop losses, it often comes too late or is inadequate to prevent serious losses and therefore tends to be disregarded. Where natural enemies of pests are destroyed by excessive use of chemicals, or by their misuse, animals not previously injurious may increase and become pests. No instances of this kind are known in beet crops in Britain, but insecticides that destroy aphid enemies (ladybird beetles, hover flies, other predatory insects and parasites) sometimes increase aphid numbers.

## CHEMICAL CONTROL

It is a mistake to think that sugar beet, or any other crop, can be grown without influencing any form of wild life. Farming itself, with or without pesticides (insecticides, fungicides, nematicides, molluscicides, rodenticides, herbicides), is 'interfering with nature' and, to grow the best and most profitable crops, pesticides may have to be used either as an insurance against a threatened attack or for the direct control of pests on the crop. Probably more pesticide is used than is essential. Some of the excesses may arise from sales pressure and some from ignorance, but it is often difficult to decide whether treatment is essential or not; when there is doubt, the grower may be justified in treating his crop as an insurance against possible heavy financial loss. It is important to ensure that the right materials are used at the right times and in the right amounts to prevent damage to the crop. It is not always necessary to kill the pests, e.g., seed treatment with dieldrin largely prevents wireworm damage but is unlikely to kill many wireworms. Large quantities of pesticides are needed to kill pests that live and feed deep in the soil; much less is needed to kill those on the soil surface or those that feed on foliage, because they are easily reached. The object should be to use the least pesticide required to obtain satisfactory control of damage, which does not necessarily imply complete control of the pest.

When sugar beet crops are treated with authorized pesticides at the rates, times and frequencies approved under the Government's Pesticides Safety Precautions Scheme, there is no risk of poisoning stock fed on tops, or of harmful residues reaching the sugar factories and perhaps contaminating beet pulp or molasses. The process of sugar extraction ensures that no undesirable residues pass into the crystalline sugar. The British Sugar Corporation, aware of the possible danger from the incorrect use of pesticides, ensures that the beet crop is not contaminated with unauthorized chemicals by prohibiting the use of any agricultural chemicals on the beet crop, or on the land during the six months before sowing, except those chemicals cleared for safe use on beet under the Pesticides Safety Precautions Scheme (Beet Growers' Contract, Clause 17—'Use of Pesticides'). Furthermore, the use of these authorized chemicals is restricted to the rates, frequencies and times approved under the Scheme. A list of authorized chemicals is published in the March and September issues of the British Sugar Beet Review.

## PESTICIDES

Before World War II satisfactory pesticides were few. Against insects there were the stomach poisons such as lead arsenate and Paris green, contact poisons such as derris and pyrethrum, and nicotine which was used both as a fumigant and a contact poison; of these only Paris green, in poison bait, is still used occasionally on beet. Many new and powerful insecticides have come into use since the war, but some that were too poisonous or too persistent have been abandoned. Those now used commercially on beet* fall into three chemical groups:

1. Organochlorine compounds:
   DDT, gamma-BHC, dieldrin.

2. Organophosphorus compounds:
   (a) Without systemic action—malathion, trichlorphon;
   (b) With systemic action—demephion, demeton-S-methyl, dimethoate, disulfoton, ethoate-methyl, formothion, menazon, oxy-demeton-methyl, phorate, phosphamidon, thiometon.

3. Carbamate compounds:
   methiocarb, (others are being tested).

Organochlorine compounds are powerful and persistent contact poisons. DDT can control many foliage-feeding insect pests but should *not* be used against aphids. Gamma-BHC is a useful soil insecticide but is liable to taint crops like potatoes and carrots up to three years after excessive applications. The use of dieldrin is restricted because of residue problems and risks to wild life; it is now used only as a seed dressing.

Some organophosphorus compounds are very poisonous to man and domestic animals, as well as to insects, but they seem generally less hazardous to wild life than the organochlorine compounds because they are less persistent and usually more selective than the latter. Malathion and trichlorphon kill mainly by contact action but can penetrate foliage locally. The remainder act systemically as well as by contact and sometimes by fumigant action: they are absorbed by plants and carried in the sap stream, so that they are particularly effective against sucking insects such as aphids. Nearly all the organophosphorus compounds listed above persist in the plant or soil for a few weeks only. Disulfoton and phorate are formulated as granules and, because the granules release the active ingredient slowly, they may persist in the soil for three to four months. The only material in this group used as a seed dressing is menazon, which controls aphids infesting seedlings until about the four rough-leaf stage when applied to raw seed; it cannot be used on pelleted seed.

**WARNING. Organophosphorus insecticides are dangerous poisons if swallowed or inhaled. They may also be absorbed into the body through the skin from splashes or from contaminated clothing, especially when concentrates are being handled. No insecticide should be stored near foodstuffs. All used containers should be**

---

*Common names of pesticides are used throughout this Bulletin. Proprietary names o currently approved products may be found in the List of Approved Products published annually and obtainable free from the Regional and Divisional Offices of the Ministry of Agriculture, Fisheries and Food or from the Ministry's Publications Branch, Tolcarne Drive, Pinner, Middlesex HA5 2DT.

**burned or buried. Unused spray and the washings from spray tanks or other equipment should be run into pits from which they can soak away into the subsoil; they should never be allowed to enter ponds, streams or rivers.** The Agriculture (Poisonous Substances) Regulations made under the Agriculture (Poisonous Substances) Act, 1952, apply to employers and workers who use certain poisonous pesticides. These regulations have been summarized in the Ministry's leaflet APS/1. Copies are obtainable free from the Ministry of Agriculture, Fisheries and Food (Publications), Tolcarne Drive, Pinner, Middlesex HA5 2DT.

PESTICIDE FORMULATIONS

*Seed dressings.* Anti-wireworm dressings are applied by the seed merchant or during pelleting; the latter ensures good distribution and adhesion, which cannot always be assured for dust-type dressings on unpelleted seed. Pelleted seed cannot be dressed by the grower.

*Dusts* are rarely employed nowadays on sugar beet because they require special machinery for application.

*Sprays* give a uniform deposit of insecticide and nearly all growers have their own sprayers. Contact insecticides may be applied at quite low volumes but, for systemic insecticides, 20 or more gallons per acre are preferable. Sprayers also used to apply herbicides must be washed thoroughly before being used to spray insecticides.

*'Prills' and granules* are a fairly recent development and a special machine is needed to apply them. It is usually mounted on the steerage hoe so that top-dressing the foliage and inter-row cultivation are combined, or for soil application it is mounted on the drill.

*Baits and pellets.* Proprietary pellets or home-prepared baits are often used to control cutworms, leatherjackets and especially slugs. They are most suitable for small areas as hand application is usually needed, but they can be applied by machines. If the bait is home-prepared the recommended rate of active ingredient should not be exceeded or the bait may become unattractive.

*Soil fumigants.* Most nematicides (eelworm-killing chemicals) are volatile liquids which must be injected into the soil or dribbled into the bottom of the plough furrow. Some are powders, 'prills' or granules which must be mixed with the soil; these decompose giving vapours toxic to eelworms.

## Further Reading

Anon. (1970). Sugar beet cultivation. *Bull. Minist. Agric. Fish. Fd, Lond.* No. 153. H.M.S.O.

Dunning, R. A. and Winder, G. H. (1972). Some effects, especially on yield, of artificially defoliating sugar beet. *Ann. appl. Biol.* **70,** 89–98.

Edwards, C. A. and Heath, G. W. (1964). *The principles of agricultural entomology.* London, Chapman and Hall.

Jones, F. G. W. (1953). The assessment of injury by seedling pests of sugar beet. *Ann. appl. Biol.* **40,** 606–7.

JONES, F. G. W., DUNNING, R. A. and HUMPHRIES, K. P. (1955). The effects of defoliation and loss of stand upon yield of sugar beet. *Ann. appl. Biol.* **43**, 63–70.

JONES, F. G. W. and JONES, M. G. (1964). *Pests of field crops*. London, Edward Arnold.

MARTIN, H. (Ed.) (1969). *Insecticide and fungicide handbook for crop protection*. Oxford and Edinburgh, Blackwell.

MARTIN, H. (Ed). (1971). *Pesticide manual*. British Crop Protection Council.

Committee papers of the Ministry of Agriculture and Fisheries' Committee for Sugar Beet Research and Education from 1935 to the present give detailed accounts of work done on sugar beet pests. See also the annual reports of Rothamsted Experimental Station.

# Guide to the Recognition of Pest Damage

THIS section, in conjunction with the Contents and the Index, is intended to help beet growers faced with pest damage to find the most appropriate sections to refer to in the Bulletin. The table on pages 20–21 shows the months in which damage by each pest occurs.

When examining a sugar beet crop for pests, first look at the field as a whole for unevenness and then more closely at patches of poor plants. Assess all other possible causes of poor growth or injury, e.g., acidity, poor drainage, or the effects of cultivations or herbicides. Pest damage is always accentuated by poor growth of plants and inadequate seedling or plant populations.

Bare patches may be due to any of the following causes:

1. The drill having missed.
2. The seed having failed to germinate; this is unusual.
3. The seed having been eaten by field mice.
4. The young shoots having been eaten away before they could appear above the soil, or the parts of the seedlings above ground having been eaten. With both kinds of damage the root can be found and, with the latter, a stump often shows above soil.
5. Serious injury by pest or disease, causing the plants to die, wither and disappear.
6. The plants having been destroyed by bad steerage hoeing.
7. Soil or plant 'poisoning' from excessive chemical (e.g., salt) or extreme acidity.
8. Lightning damage.

Where bare patches are caused by insect or allied pests, search for them on the plants, or in the soil, around the edges of the areas attacked. Any unknown creatures found on the plants, or in the soil near the roots, should be placed in a tin with a few pieces of leaf and some soil; do not ignore them however small they may be. Look out for insects that leap, fly or drop from the seedlings when disturbed. Insects and allied pests may be sent for identification to the Entomologist at the Regional Headquarters of the Agricultural Development and Advisory Service. If damaged plants are sent, they should be fresh and packed carefully, preferably in a polythene bag.

## SEEDLINGS

A beet plant is considered to be in the seedling stage until the recommended time for singling, i.e., until it has produced two to four rough (secondary or true) leaves in addition to its two cotyledons (seed leaves). Some of the pests that are most damaging to the seedling stage continue to cause more minor damage later. Crawl or walk along the rows, bending low so as to bring the eyes near to the seedlings. Examine damaged seedlings and surrounding soil in the order given below:

### LEAVES

Examine the cotyledons and rough leaves, noting the type of damage caused or any distortion of the leaves or petioles (leaf stalks).

SOIL

Many of the insects and related pests which seriously damage beet seedlings live in the top soil within two or three inches of the surface, but some may be deeper. Watch carefully close to the seedlings for the movements of tiny, dark brown beetles about the size of a pin's head (pygmy beetle). Scrape away, a little at a time, the soil from around freshly damaged seedlings and, if necessary, continue to a depth of four to six inches with a trowel. Break up the soil gently, spreading it out on paper or on the soil surface, and search through it for wireworms, millepedes, symphylids, beetles, etc. Soil creatures that are the same colour as the soil often give themselves away by moving. Unknown creatures should be placed in a tin with a little soil.

ROOT AND STEM BELOW GROUND

Examine the underground portion of the seedling carefully, looking for discoloration, shrivelling, constrictions, tiny round black pits, and any other signs of damage. Damaged root tissue soon turns black and this shows much more clearly on washed seedlings floating in a dish of water.

| Symptoms | Causes |
|---|---|
| *Seed* | |
| Failure to germinate | Rarely caused by pests attacking true seed, although millepedes and allies may feed on the dead seed |
| Seeds dug up | Field mice |
| Seedling germinates but fails to emerge | Soil capping, too deep sowing, or pest, disease or herbicide damage before emergence |
| *Cotyledons and/or Leaves* | |
| Crushed, bruised, tattered | Implements, hail, wind |
| Pulled down into soil | Earthworms |
| Leaves mined by maggots between the two surfaces of the leaf | Beet leaf miner |
| Upper or underside of leaves mined superficially by small, dark green caterpillar which fastens leaf round itself | Tortrix moth caterpillars |
| Holes, not always penetrating both surfaces of the leaf | Beet flea beetle, tortoise beetles, pygmy beetle |
| Very small pits, not penetrating through leaf | Springtails |
| Segments cut from the leaves; stem may also be cut through at ground level | Cutworms, leatherjackets, beet carrion beetle, sand weevil, beet leaf weevil, slugs |
| Leaves more or less completely eaten away, i.e., plants 'grazed'; growing points usually intact, but only stumps may be left | Rabbit, birds, slugs |

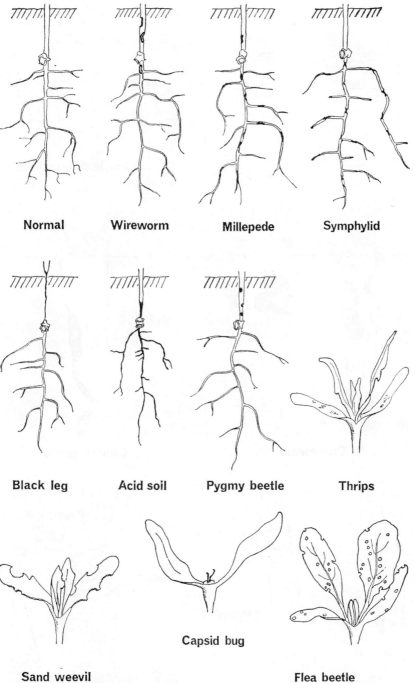

**Normal**   **Wireworm**   **Millepede**   **Symphylid**

**Black leg**   **Acid soil**   **Pygmy beetle**   **Thrips**

**Capsid bug**

**Sand weevil**
**(slug, cutworm, leatherjacket**
**and carrion beetle cause**
**similar damage)**

**Flea beetle**
**(tortoise and pygmy**
**beetles cause**
**similar damage)**

Fig. 4(a). Injury to seedling roots and foliage

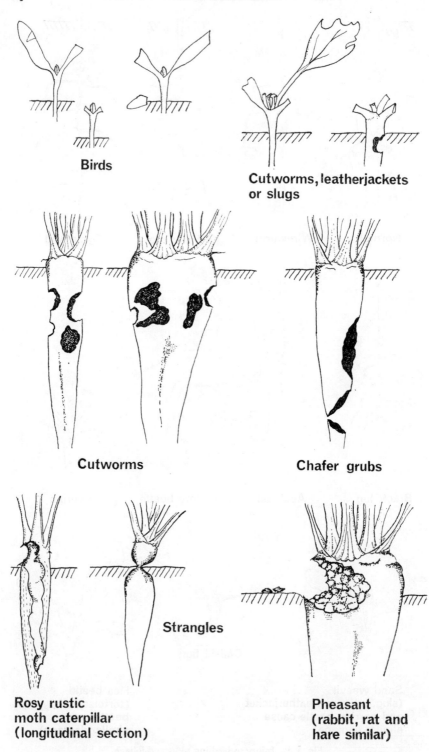

**Birds**

**Cutworms, leatherjackets or slugs**

**Cutworms**

**Chafer grubs**

**Strangles**

**Rosy rustic moth caterpillar (longitudinal section)**

**Pheasant (rabbit, rat and hare similar)**

Fig. 4(*b*). Injury to seedlings (top row) and to roots of older plants

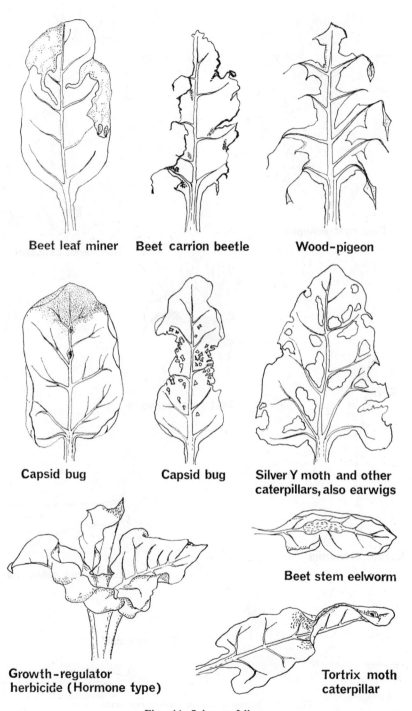

Beet leaf miner    Beet carrion beetle    Wood-pigeon

Capsid bug    Capsid bug    Silver Y moth and other caterpillars, also earwigs

Growth-regulator herbicide (Hormone type)

Beet stem eelworm

Tortrix moth caterpillar

Fig. 4(c). Injury to foliage

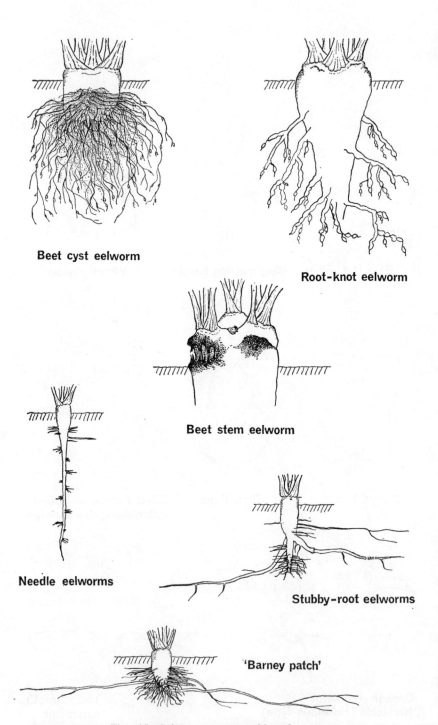

Beet cyst eelworm

Root-knot eelworm

Beet stem eelworm

Needle eelworms

Stubby-root eelworms

'Barney patch'

Fig. 4(d). Injury to roots caused by eelworms
(top three × ⅕; bottom three × ⅓)

| Symptoms | Causes |
|---|---|
| *Cotyledons and/or Leaves*—contd. | |
| Leaves curled with aphids feeding on underside | Aphids |
| Leaves wilted and, later, shrivelled | Root or stem damaged by pest, disease or herbicide |
| Heart leaves narrowed, sometimes distorted, and with blackened, roughened edges | Thrips |
| Galls on petioles; leaves thickened and distorted | Stem eelworm |
| Dead growing point, cotyledons enlarged, and multiple crown developing later | Stem eelworm, capsids, mechanical injury |
| *Plants distorted: leaf edges crinkled, or leaf stalks fused, or leaves loosely fused together by growth and not by caterpillar web | Spray drift or soil residue of herbicide |
| *Diffuse patches of stunted and healthy plants intermingled, on sandy soil—Docking disorder | Stubby-root eelworms, needle eelworms |
| *Distinct patches of stunted plants in otherwise healthy crop on loamy, alkaline soil—Barney patch | Unknown (see page 88) |
| *Stem and Root* | |
| Split in skin of root above seed level | Normal secondary thickening of root |
| Small, round, black pits on surface, usually above seed level | Pygmy beetle |
| Browned or blackened areas, wounds obvious | Wireworms, chafer grubs, cutworms, leatherjackets, bibionid larvae, slugs, swift moth caterpillars |
| Browned or blackened areas, wounds much less obvious | Millepedes, symphylids, springtails |
| Root tunnelled by flesh-coloured caterpillar | Rosy rustic moth caterpillar |
| Seedlings uprooted | Birds, usually rooks, searching for insects |

## PLANTS AFTER SINGLING

LEAVES AND CROWN

Examine the outer, middle and heart leaves, and note any insects present and the type of damage. Also examine the crown for damage, canker, or the 'multiple crown' condition.

---

*May become apparent by singling time, but more pronounced later.

SOIL

Look on the surface of the soil around the plant for any creatures which may have fallen from the leaves, and then search the surface soil by scraping away a little at a time down to about three inches. Dig up the plant and the surrounding soil down to a depth of nine inches. Place on a sack or on a clear patch of soil and break up the soil gently, catching any creatures which are found and placing them in a tin with a little moist soil.

ROOTS

Examine the root for signs of damage or discoloration. Cut it through, noting any signs of decay or of insect tunnels. Note the state of the side roots and the extent of fanginess.

| SYMPTOMS | CAUSES |
|---|---|
| *Leaves* | |
| Outer leaves yellowed and wilting | Nitrogen deficiency, drought, senescence |
| Leaves with yellow speckles later becoming necrotic; upright, triangular, edges curled towards upper surface—speckled yellows | Manganese deficiency |
| Leaves with bright yellow blotches | Tobacco rattle virus |
| Mottled yellowing between green vein tissue | Iron deficiency (occasional plants on chalk soil), beet mosaic virus |
| Yellowing along veins | Excess uptake of the herbicide lenacil |
| Diffuse yellow or orange colouring, leaves thickened and brittle | Beet yellows virus, beet mild yellowing virus, downy mildew |
| Yellowing extending from leaf edges, which become blackened | Magnesium deficiency |
| Yellowing confined mainly to the end portion of the leaf blade, with stab marks on veins and distorted growth | Capsids |
| Yellowing in late summer after long spell of hot, dry weather; mites, eggs and webbing on underside of leaves | Red spider mite |
| Leaves curled towards the undersides, which are covered with black aphids | Black bean aphid |
| Leaf veins twisted, leaves puckered | Cuckoo-spit insect, capsids |
| Leaves with many small punctures, sometimes with ragged, yellow or brown-edged holes | Capsids |
| Leaves mined by maggots between the two surfaces of the leaf | Beet leaf miner |
| Upper or underside of leaves mined superficially by small, dark green caterpillar which fastens leaf round itself | Tortrix moth caterpillars |

| Symptoms | Causes |
|---|---|
| Underside of leaves mined superficially by small green caterpillar which does not bind leaves together | Diamond-back moth caterpillar |
| Holes, not always penetrating to the other surface | Beet flea beetle, tortoise beetles |
| Small to large holes, the latter sometimes fusing | Pygmy beetle (which feeds only on heart leaves), beet carrion beetle, tortoise beetles, earwigs, silver Y moth or other caterpillars, slugs |
| Pieces torn from leaf, sometimes leaving only mid-rib and main veins | Birds (especially wood-pigeon), hailstones |
| Leaves and petioles partly eaten, torn from crown, scattered | Rabbit, hare, coypu |
| Leaves wilted or flagging; flesh-coloured caterpillar tunnelling in crown | Rosy rustic moth caterpillar |
| Parts, especially tips, of leaves wilted | Capsids |
| Wilted or flagging, no obvious superficial cause | Drought, beet cyst eelworm, root injury or disease from various pests or diseases |
| Tops distorted, leaf stalks fused, or leaf edges crinkled | Herbicide spray drift or soil residue |
| Leaves silvered extensively on underside | Sun and wind |
| Diffuse patches of stunted and healthy plants intermingled, on sandy soil—Docking disorder | Stubby-root eelworms, needle eelworms |
| Sharply defined patches of stunted plants in otherwise healthy crop on loamy, alkaline soil—Barney patch | Unknown (see page 88) |

*Crown*

| | |
|---|---|
| Cankered; may or may not be associated with 'multiple crown' condition | Stem eelworm, boron deficiency |
| Eaten into, and sometimes through, at or just above soil level | Pheasant, rabbit, hare, coypu, rat |

*Roots*

| | |
|---|---|
| Tunnelled | Rosy rustic moth caterpillar |
| Stunted and sometimes fanged | Soil conditions, acidity, disease or mechanical injury, stubby-root eelworms, needle eelworms |
| Very stunted and very fangy | Causes of Barney patch |
| Profusion of living and dead rootlets: | |
| (i) with numerous white or brown lemon-shaped cysts adhering | Beet cyst eelworm |

| SYMPTOMS | CAUSES |
|---|---|
| (ii) without cysts, but roots fangy, the fangs often horizontal | Stubby-root eelworms; causes of Barney patch |
| Knot-like galls on rootlets | Root-knot eelworms |
| Roughly circular, black pits scattered over surface of root and extending to six inches below soil level: | |
| (i) minute pits | Pygmy beetle |
| (ii) small pits | Wireworms |
| Large, irregular pits eaten in the roots, small tap roots may be eaten through: | |
| (i) usually just below soil level | Cutworms, leatherjackets, swift moth caterpillars |
| (ii) usually deeper | Chafer grubs |
| Constriction of root at or below soil level; plant snaps off some time after singling—strangles | Exposure of root to wind damage at singling |

The months in which the pests of beet may be found feeding on the crop (*) and those months in which most damage is caused (**) are shown below, also the site of damage for each pest.

Months found on crop

| | Apr. | May | June | July | Aug. | Sept. | Site of feeding |
|---|---|---|---|---|---|---|---|
| **I.** *Insect and allied pests* | | | | | | | |
| *(a) Mainly root-eating pests* | | | | | | | |
| Wireworms . | ** | ** | ** | * | * | * | { Mainly above seed level. |
| Bibionid larvae. | ** | ** | * | | | | Below { Above and below seed level. |
| Millepedes . | ** | ** | ** | * | * | * | soil only |
| Symphylids . | ** | ** | * | * | * | * | { Mainly below seed level. |
| Pygmy beetle . | ** | ** | ** | * | * | * | Mainly below, but sometimes above, soil level. |
| Springtails . | ** | ** | * | * | * | * | Above and below soil level. |
| Earthworms . | ** | ** | * | * | * | * | |
| Cutworms . | ** | ** | ** | ** | * | * | Below, and often above, soil level. |
| Leatherjackets . | ** | ** | ** | * | | | |
| Slugs . . | ** | ** | * | * | * | * | |
| Chafer grubs . | * | ** | ** | ** | ** | * | Below soil level only. |
| Swift moth caterpillars . | ** | ** | * | | | | |
| Rosy rustic moth caterpillar . | * | ** | ** | * | | | In the crown and tap root. |
| *(b) Leaf-eating pests* | | | | | | | |
| Beet flea beetle. | ** | ** | * | * | * | * | On leaf, mainly the upper surface. |
| Sand weevil . | ** | ** | * | | | | On leaf edges. |
| Beet carrion beetle . | ** | ** | ** | * | | | |

Months found on crop

| | Apr. | May | June | July | Aug. | Sept. | Site of feeding |
|---|---|---|---|---|---|---|---|
| Beet leaf miner. | | ** | ** | ** | * | * | Inside leaf, between upper and lower surfaces. |
| Tortrix moth caterpillars . | | ** | ** | | | | Leaf surface, after binding leaf or leaves together. |
| Diamond-back moth caterpillar . | | | | ** | | | Under surface of leaf, mining when small. |
| Silver Y moth caterpillar . | | | * | ** | ** | * | Leaf surface and edge. |
| Earwig . . | | | ** | * | * | * | |
| Tortoise beetles | ** | ** | * | * | * | * | |

(c) *Sap-sucking pests*

| | Apr. | May | June | July | Aug. | Sept. | Site of feeding |
|---|---|---|---|---|---|---|---|
| Black bean aphid | | ** | ** | ** | * | * | Leaves, mainly under-side and on or near veins. |
| Peach-potato aphid . . | ** | ** | ** | ** | * | * | |
| Other aphids . | * | * | * | * | | | |
| Capsids . . | ** | ** | ** | ** | * | * | Growing point of seedlings, leaves (especially veins) and leaf stalks at any time. |
| Thrips . . | ** | ** | * | * | * | * | Leaf surface. |
| Leafhoppers . | | * | * | * | * | * | |
| Red spider mite | | | | * | * | * | |

II. *Eelworm pests*

| | Apr. | May | June | July | Aug. | Sept. | Site of feeding |
|---|---|---|---|---|---|---|---|
| Beet cyst eel-worm . | * | ** | ** | ** | * | * | In the rootlets. |
| Root-knot eel-worms . | * | ** | ** | * | * | * | In galls on the rootlets. |
| Stem eelworm . | ** | ** | * | * | * | ** | In leaves, leaf stalks, growing point (of seedlings) and the root crown. |
| Free-living eel-worms . . | ** | ** | ** | ** | * | * | On the primary tap root (seedling only) and root tips. |

III. *Bird and Mammal pests*

| | Apr. | May | June | July | Aug. | Sept. | Site of feeding |
|---|---|---|---|---|---|---|---|
| Rook . . | * | ** | * | | | | Uproots seedlings searching for insects. |
| Wood-pigeon . | * | * | ** | ** | * | * | Leaves. |
| Game birds . | ** | ** | ** | * | * | * | Leaves and, after singling, crowns. |
| Other birds . | ** | ** | * | | | | Leaves. |
| Field mice . | ** | ** | | | | | Seeds dug up. |
| Rabbit . . | * | ** | ** | * | * | * | Leaves, petioles and crowns. |
| Hare . . | * | ** | ** | ** | * | * | |
| Coypu . . | * | ** | ** | ** | ** | * | |

\* Feeding occurring.

\*\* Main damage occurring.

# FURTHER READING

ANON. (1968). *Les principaux parasites et maladies de la betterave sucrière*. Paris, Institut Technique Français de la Betterave Industrielle.

DUNNING, R. A. (1972). Sugar beet pest and disease incidence and damage, and pesticide usage, in Europe: report of an I.I.R.B. enquiry. *J. int. Inst. Sug. Beet Res*. (In the press)

ERNOULD, L. and VAN STEYVOORT, L. (1958). *Atlas des ennemis et maladies de la betterave*. Tirlemont, Institut Belge pour l'Amélioration de la Betterave.

HULL, R. (1960). Sugar beet diseases. *Bull. Minist. Agric. Fish. Fd, Lond*. No. 142. H.M.S.O.

LÜDECKE, H. and WINNER, C. (1966). *Farbtafelatlas der Krankheiten und Schädigungen der Zuckerrübe*. Frankfurt am Main, DLG-Verlag. (Includes coloured illustrations with English captions, of pests and pest damage.)

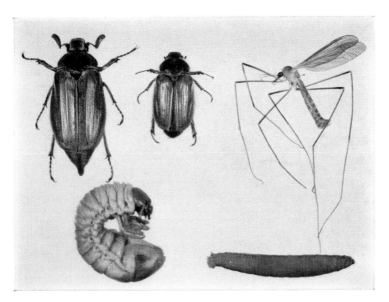

Cockchafer          Summer chafer          Crane fly

Cockchafer grub                 Leatherjacket

Silver Y moth                          Turnip moth

Silver Y moth                    Turnip moth caterpillar
caterpillar                         or cutworm

Silver Y moth pupa                    Turnip moth pupa

Swift moth                        Swift moth caterpillar

LARGE INSECTS INJURIOUS TO SUGAR BEET IN THE LARVAL STAGE
NATURAL SIZE

PLATE I

| Click beetle | Wireworm | Sand weevil | Beet carrion beetle | Beet carrion beetle larva |
| | | Capsid bug | | |

Beef leaf weevil          Potato stem borer          Millepede (*Julus* sp.)

Tortoise beetle          Froghopper          Mangold fly

Pygmy beetle          Beet leaf bug          Mangold fly larva

          Mangold fly puparium

Polygonum leaf beetle          Leafhopper or Jassid          Beet flea beetle

INSECTS INJURIOUS TO SUGAR BEET. ENLARGED

PLATE II (upper × 1½, lower × 2)

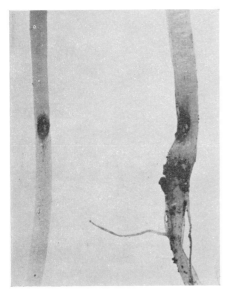

Pygmy beetle damage to root of seedling: characteristic small, oval, blackened pits, usually above seed level.

Symphylid damage to seedling root.

Caterpillar damage to beet leaves in late summer.

Slug damage: note typical irregularity, also cloddy soil.

PLATE III

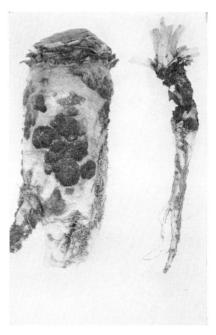

Seedling injured by the feeding of the beet flea beetle.

Injury caused by a heavy cutworm attack in July. Backward plants like that on the right may be completely severed just below ground level.

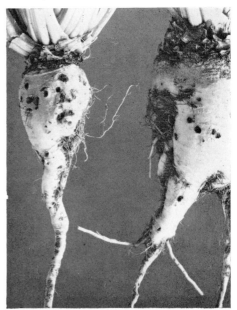

Wireworm-damaged beet seedlings. The jagged, blackened wounds are almost always between seed and soil level.

Wireworm pits on the surface of mature sugar beet. This type of injury is of little importance.

PLATE IV

# Insect and Allied Pests

ADULT insects can usually be recognized by their hard external skeleton or cuticle, the division of the body into head, thorax and abdomen, and the possession of three pairs of walking legs and, usually, wings. The young or larval stages of some species resemble the adults, but others are grub-like without obvious appendages and are very different from the adults; species in which the larvae differ from the adults have an intervening stage, the pupa (or chrysalis), during which the transformation occurs. Millepedes, symphylids and mites resemble insects in having a hard external skeleton, but they are wingless, have more than three pairs of legs in the adult stage, and the body is not divided into head, thorax and abdomen. Slugs are soft-bodied and easily recognized.

Insects are cold-blooded, that is they do not generate heat or control their body temperature, which fluctuates with the air or soil temperature. The winter therefore makes them torpid and greatly slows or stops their movement, feeding and multiplication. The winter is passed as adult, egg, larva or pupa according to species, either in soil, hedgerows, copses or plant litter, or sometimes on host plants in sheltered spots. How pests overwinter and the numbers that survive often influence the amount of injury caused to ensuing crops. Even where new beet crops are not infested at sowing time because the insects overwinter elsewhere, the power of flight possessed by most insects enables them to invade and establish themselves in the young crop.

The mouthparts of insects and allied pests are of two kinds: biting, or piercing and sucking. Species in which both adults and larvae have biting mouthparts are considered first; they eat parts of the plant and so cause obvious wounds, such as areas cut from leaves, pits in roots and stems, tunnels or mines. They are considered in two groups depending on whether they are mainly root-eating or entirely leaf-eating pests (see list on pages 20–21).

## MAINLY ROOT-EATING PESTS
### WIREWORMS*

Three common species of wireworm attack sugar beet, *Agriotes lineatus* (L.), *A. obscurus* (L.) and *A. sputator* (L.), and all are very similar in appearance, life history and habits. The adult stage (Plate II) is a dull brown insect, ¾ to ⅝ in. long and called the click beetle or skipjack because, when placed on its back, it leaps into the air with a clicking noise. Click beetles are found in grass or running actively on the soil surface from April to July. After mating, the female beetles lay eggs in the upper layers of the soil and these hatch in three to four weeks to give minute wireworms. The newly emerged wireworms are barely 1/20 in. long and are rarely seen; the wireworms usually seen are from two to five years old and ½ to 1 in. or more long. They have tough yellow skins, a pair of powerful jaws, and three pairs of tiny legs immediately behind the head. When fully grown, they burrow deeper into the soil, make a small earthen cell and transform into the pupal stage. This transformation occurs in the late summer of the fourth or fifth year; the

---

* See also Advisory Leaflet 199, available from the Ministry (p. 108).

23

C

adult click beetle emerges from the pupa the same autumn, overwinters in the soil and comes to the surface in the spring. Fig. 5 shows the life cycle.

PLANT INJURY

There are two periods of active feeding by wireworms during the year: one in the spring when their feeding coincides with the critical seedling stage in many crops, and another in the autumn. Wireworm injury takes different forms in different crops; in sugar beet it shows as small wounds that soon blacken on the seedling stem below ground level (Plate IV and Fig. 4(a)). The wound is small, for wireworms eat much less than cutworms or leather-jackets, but is usually enough to make the seedling wilt and ultimately die.

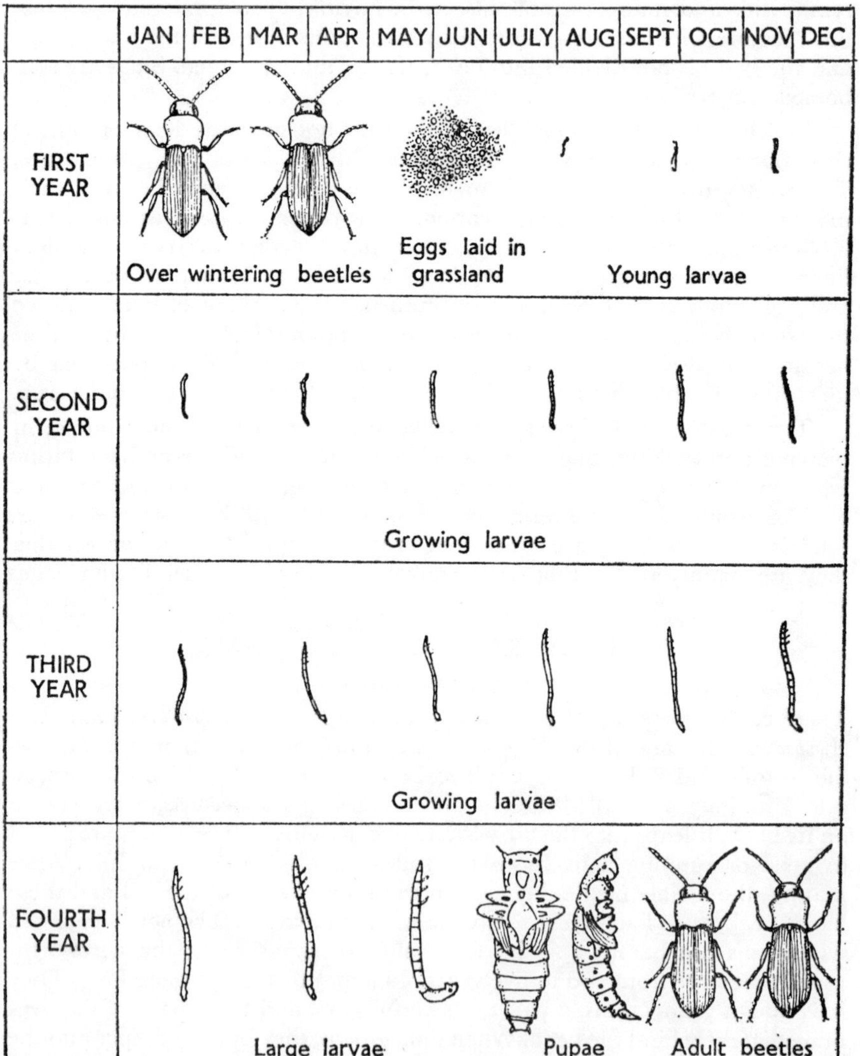

| | JAN | FEB | MAR | APR | MAY | JUN | JULY | AUG | SEPT | OCT | NOV | DEC |
|---|---|---|---|---|---|---|---|---|---|---|---|---|
| FIRST YEAR | Over wintering beetles | | | | Eggs laid in grassland | | | | Young larvae | | | |
| SECOND YEAR | Growing larvae | | | | | | | | | | | |
| THIRD YEAR | Growing larvae | | | | | | | | | | | |
| FOURTH YEAR | Large larvae | | | | | | Pupae | | | Adult beetles | | |

Fig. 5. Life cycle of a click beetle

(Reproduced by courtesy of Edward Arnold from *Pests of Field Crops* by F. G. W. Jones and M. G. Jones, 1964)

Sugar beet cannot withstand attack until after the singling stage, and even larger plants that survive damage may be so crippled that they remain stunted. Injury to fully grown roots takes the form of small, blackened pits where feeding occurred below ground level (Plate IV). This injury occurs during the period of autumn feeding and does little harm because the wireworms do not burrow into the flesh of the root as they do into potato tubers.

INCIDENCE OF ATTACK

Most permanent grassland harbours large populations. When the grass is ploughed and the turf decomposes, the wireworms turn to the arable crops for food and, because of their long life cycle, can injure crops for about three years, usually causing most damage in the second year. The problem after ploughing up grassland is to decide whether sugar beet can be grown safely and whether protective measures are needed to ensure its survival. Arable land usually contains too few wireworms to cause trouble, but some areas have an 'arable wireworm' problem; this rarely leads to large losses, although the stand of plants may be disappointing and irregular. Such attacks are usually sporadic and cannot be anticipated. Short leys do not cause great increases in the wireworm population unless they follow soon after permanent grass, but they tend to aggravate the 'arable wireworm' problem. Sainfoin, ryegrass and cereals favour wireworms more than do lucerne, red clover or trefoil. Wireworm populations much greater than the intended plant population (30,000 per acre) are common in arable land, but usually the wireworms either do not feed or fail to find beet plants. Any soil or climatic factors favouring wireworm activity and retarding the growth of the seedlings will increase plant losses. Sugar beet grown in chalky soil seems especially liable to be harmed by small wireworm populations. Wireworms were one of the most important pests of beet seedlings, but the annual loss declined sharply after 1953 because of decreased ploughing of old grass, the use of insecticide seed-dressings on beet and other crops, and the use of insecticides as a soil treatment for some crop pests. The current trend towards drilling seed to a stand (e.g., 5-in. spacing, equivalent to 60,000 seed places per acre) is likely to increase the importance of small wireworm populations (see page 4).

CULTURAL CONTROL

During and immediately after World War II, whenever fresh grassland was broken, it was customary to have the wireworm population of the soil estimated by the National Agricultural and Advisory Service, and from this to decide which crops could safely be grown and whether treatment with insecticide was necessary. Crops susceptible to wireworm injury are sugar beet, mangels, swedes, carrots, maincrop potatoes and spring cereals; more resistant crops include winter wheat and rye, winter and spring beans, peas, linseed, mustard, silage and grazing mixtures.

After ploughing long ley or permanent pasture, sowing beet at the small seed rates now used is inadvisable unless the crop is protected by treating the soil with insecticide. After grass, every effort should be made to obtain a good and adequately consolidated seedbed and any deficiencies in lime, phosphate or potash must be corrected. The seed rate should be greater than usual and seed should not be sown very early (i.e., in March) because then growth is often slow at first; with later sowings the seedlings pass more rapidly through the susceptible stage. The wireworm population declines under arable crops and it is usually safe to grow sugar beet in the third year, unless crops in the first and second year were severely attacked. A full fallow, the land being broken in February or March, decreases wireworms to about 5 per cent of their original numbers. Ploughing later is far less beneficial: April ploughing and then fallowing halves the population and is similar in effect to growing an arable crop, but July ploughing kills only the smallest wireworms and leaves the large ones unaffected.

CHEMICAL CONTROL

All sugar beet seed, whether or not pelleted, is treated with an insecticidal dust dressing; the growers' packs of seed bear the label 'anti-wireworm dressed'. The dressing is cheap and convenient; it deters wireworms from feeding on beet seedlings and largely prevents seedling losses from the wireworms that usually occur in arable land, but it does little to decrease wireworm numbers. Dieldrin (9 oz of 40 per cent dressing per cwt of seed) is used at present; the amount of insecticide applied per acre is very small (about 1/15 oz on 2 lb of seed, or 2 parts in a thousand million parts of soil), so there is little risk to wild life. However, current research work aims at substituting dieldrin with a less persistent chemical which is also effective against other seedling pests, especially pygmy beetle.

Where the wireworm population exceeds 250,000 per acre and close seed spacing and/or late drilling is impractical, the soil should be treated with gamma-BHC. Apply 8 oz of active ingredient (10 fl. oz of 80 per cent suspension) in at least 20 gal of water per acre and work lightly into the seedbed. If drilling less than 2 lb seed per acre (5 in. spacing or wider), then soil treatment may well be advisable where there are only 100,000 wireworms per acre, especially where genetic monogerm seed is drilled early.

When, unexpectedly, wireworms are found to be damaging seedlings in April or May, partial protection can be provided by gamma-BHC. Apply 12–16 oz of active ingredient (15–20 fl. oz of 80 per cent suspension) in as large a volume of water per acre as convenient, and then steerage hoe as close as possible to the rows to help mix the insecticide into the surface soil. Less active ingredient is needed when applied by band spraying. Potatoes or carrots should not be grown for at least two years after a soil treatment with gamma-BHC, because of the risk of taint.

Further work is in progress to improve methods of protecting the seedling populations obtained by current and likely future methods of drilling.

FURTHER READING

DUNNING, R. A. and WINDER, G. H. (1965). Sugar beet seedling populations and protection from wireworm injury. *Proc. 3rd Br. Insecticide & Fungicide Conf. Brighton, 1965.* 88–99.

JONES, D. P. and JONES, F. G. W. (1947). Wireworms and the sugar-beet crop: field trials and observations. *Ann. appl. Biol.* **34,** 562–74.

STAPLEY, J. H. (1949). *Pests of farm crops.* London, Spon.

## BIBIONID LARVAE

Larvae of bibionids occasionally damage seedling beet slightly in April and May. The larvae resemble small leatherjackets (see page 33) but have a well-developed head and twelve legless segments, each with a characteristic band of fleshy protuberances. Several species probably attack beet; the larvae of some species are gregarious. Damage resembles that caused by wireworms but no control measures can be recommended.

FURTHER READING

MORRIS, H. M. (1921). The larval and pupal stages of the Bibionidae. *Bull. ent. Res.* **12,** 221–32.

## MILLEPEDES*

Crop damage by millepedes has increased since 1964 and occurs most often on the heaviest silt and peat soils; those which have much ploughed-in stubble and other organic matter, and are moist and open-textured, seem preferable for millepedes, symphylids, springtails and pygmy beetle, which are often found together in suitable soils.

Millepedes move actively and are most readily found near the soil surface in May and June, but the flat millepede is active there also in the autumn; when the soil is dry all species move much deeper.

Two species cause damage: the flat millepede (*Brachydesmus superus* Latz.) and the spotted snake millepede (*Blaniulus guttulatus* (Bosc)) (Plate VII). The former has a light brown flattened body up to ½ in. long; the latter is up to ¾ in. long, translucent to creamy-yellow with a row of red spots along each side, and forms a tight coil when disturbed. Species of other genera occasionally damage beet in England. The young of flat millepedes, but not other species, are readily found in the soil and can be confused with *Onychiurus* springtails (page 31), since they are white, sluggish and have three pairs of legs. In all millepedes growth is accompanied by increase in number of segments, each of which bears two pairs of legs.

PLANT INJURY

Severe damage to young seedlings causes wilting and death, but usually only stunts older seedlings. In fields where sugar beet is attacked, the spotted snake millepede can be extremely numerous, up to ten million per acre, but the flat millepede is rarely as numerous. Spotted snake millepedes aggregate and coil round the base of the hypocotyl (region between root and stem) at or below seed level, where they injure the skin and expose the underlying tissues, which become discoloured (Fig. 4(a)); flat millepedes usually feed above seed level. Millepedes tend to feed on roots which have injured surfaces exuding sap, for instance plants previously injured by wireworms, symphylids and the ubiquitous pygmy beetle. Therefore, it may sometimes be the injury which attracts the millepedes to the plants, leading to further damage. However, they are sometimes the primary cause of injury, damaging plants from germination until the four rough-leaf stage. Millepedes are a serious pest of seedling beet in Belgium and France, where herbicides and wide seed-spacing are blamed, and much research is being done on their biology and control.

CONTROL

It is difficult to develop control measures for millepedes because they are so sporadic, almost exclusively soil-inhabiting and usually occur with other

---

* See also Advisory Leaflet 150, available from the Ministry (p. 108).

soil pests. Current research work suggests that modification to soil husbandry in fields at risk may be well worth while. Gamma-BHC at 16 oz of active ingredient (20 fl. oz of 80 per cent suspension) per acre, worked into the seedbed, gives partial control. The standard insecticidal seed treatment may give some protection, but no treatment can confidently be recommended to control damage when millepedes have started to attack seedlings. Gamma-BHC sprayed at 16 oz of active ingredient (20 fl. oz of 80 per cent suspension) per acre concentrated in a band along the rows is sometimes beneficial, especially with the flat millepede when heavy rain follows treatment. However, such a treatment is only worth trying if the crop is in danger of being destroyed. Potatoes or carrots cannot be grown for at least two years after application of gamma-BHC, because of the risk of taint.

FURTHER READING

BLOWER, J. G. (1958). British Millipedes (Diplopoda). *Linn. Soc., Lond. Synopses of British Fauna* No. 11.

BRENY, R. and BIERNAUX, J. (1966). Diplopodes belges: position systématique et biotypes. *Communication présentée à la réunion de la Société Royale d'Entomologie de Belgique, Bruxelles, January 1966.*

STEPHENSON, J. W. (1960). The biology of *Brachydesmus superus* (Latz.) Diplopoda. *Ann. Mag. nat. Hist.* (13th series) **3,** 311–9.

## SYMPHYLIDS*

Symphylids are sometimes found feeding at the roots of sugar beet seedlings. They are not insects but are related to them and, more closely, to centipedes and millepedes, having many pairs of legs. The species that damages beet is the glasshouse symphylid, *Scutigerella immaculata* (Newp.), so called because it is a common pest of glasshouse crops; it also occasionally damages potatoes in the field (Plate VII). The same species is troublesome on beet and other field crops in the U.S.A.

Symphylids are white and extremely active, with a pair of long, rapidly palpating antennae on their head. When fully grown they are about ¼ in. long with twelve pairs of legs; the young have six pairs of legs. Symphylids are rather delicate creatures and can only move in the soil via cracks and crevices, which probably accounts for them being a problem on the moister and heavier soils, where they are often found with millepedes and spring-tails. Recent surveys have shown them to be prevalent also on limestone soils. They rarely feed above seed level and are often found much deeper; because of their habits they are not often noticed in the soil. If samples of soil are stirred in a bucket of water, symphylids, springtails (see page 31) and some millepedes float on the surface.

By their feeding, symphylids can cause seedlings to wilt and die. Damaged areas are irregular in extent and very soon blacken (Plate III and Fig. 4(a)). The damage is virtually indistinguishable from that caused by millepedes, and the two pests are often found together. Less severe injury stunts seedling growth; the larger and more vigorous the seedlings, the better they are able to withstand attack.

---

* See also Advisory Leaflet 484, available from the Ministry (p. 108).

No recommendations can be made at present for the prevention of symphylid damage in beet fields. Damage cannot be forecast and the pest cannot be reached by any chemical treatment once damage is occurring. If re-drilling is necessary, then soil treatment with gamma-BHC as recommended for wireworm control (see page 26) may be advisable.

FURTHER READING

EDWARDS, C. A. (1961). The ecology of Symphyla: Part III. Factors controlling soil distributions. *Entomologia exp. appl.* **4,** 239—56.

MICHELBACHER, A. E. (1938). The biology of the garden centipede, *Scutigerella immaculata. Hilgardia* **11,** 55–148.

## Pygmy Beetle

The pygmy beetle or pygmy mangold beetle (*Atomaria linearis* Steph.) is a brown beetle, about $\frac{1}{16}$ in. long (Plate II). Before 1935 it was the most important pest of sugar beet and was the only pest of any consequence on the best silts; its depredations were the result of growing beet after beet. After 1935 its importance diminished because a clause was included in factory contracts forbidding the growing of beet after beet or mangels, a restriction meant primarily to control the increase of beet cyst eelworm (see page 78).

The pygmy beetle occurs in all beet-growing areas and can usually be found in any beet or mangel field if sought diligently. Large resident populations exist wherever beet is grown intensively but, because of its small size, it is often overlooked and the injury it causes attributed to other pests. The beetle takes a toll of seedlings in the main beet-growing areas and this can be serious where seeds are sown at wide spacings.

PLANT INJURY

The beetle starts to feed in March and continues well into the summer, chiefly below ground, but also above ground in showery weather. It makes characteristic pits in the hypocotyl and roots, or small circular holes in the cotyledons and heart leaves (Plates III, VI and Fig. 4(a)). Leaf injury is often symmetrical because it occurs when the leaves are folded in the terminal bud, but this is less serious than bites below ground in the young hypocotyl. The effect on the crop depends on the number of beetles and the stage of growth of the plants. When many beetles are present, every seedling has numerous bites and many wither and die. Once past the cotyledon stage, attacked plants usually survive, although some are crippled and produce mis-shapen roots at harvest. Fresh bites can still be found in the tap root at a depth of 1–6 in. as late as July. Beetles sometimes feed in the terminal shoots of seed crops but such injury is quite unimportant. Mangels, red beet, fodder beet and spinach are also attacked. Beetles can be induced to feed on many weeds in the laboratory but pay little attention to them in the field. In beet crops almost all the beetles assemble round the beet plants, although a few are found adjacent to fat hen (*Chenopodium album* L.).

LIFE CYCLE

Eggs are laid in the soil from April onwards and the larvae feed on the roots of sugar beet. The earliest adults to emerge mature in about three weeks and begin to lay eggs from which a second generation of beetles arises. Because egg laying is protracted, new beetles continue to emerge until the following spring and accumulate in great numbers in the old beet fields from autumn onwards. New beetles are pale at first but soon darken. A few disperse in the autumn but most spend the winter in the soil of old beet fields where numbers may exceed a million per acre; other fields are almost free from beetles in the following spring. Mass flights from the old beet fields commence with the first spells of fine, spring weather when the air temperature approaches 70°F (21 °C); the first flights are usually in mid-April but, if the weather continues cold, may be delayed until June. Flights continue to occur on suitable occasions in May and June but decrease from July onwards. After the start of dispersal flights, beetles and their characteristic injury begin to appear in the new beet fields; numbers of beetles may increase to 500,000 per acre shortly after singling. It is fortunate that they do not transmit viruses because, before dispersing, many feed on old beet crowns which have survived the winter and which are sometimes infected with viruses.

The method of overwintering shows the reason why complete loss of crop is likely only where beet follows beet or other host crops. In such circumstances the beetles begin to feed as soon as the seed germinates and the crop may disappear very quickly. If, by error, there is any overlapping of cropping in part-cropped fields, the area where seedlings are destroyed often coincides within a foot or so with the site of the previous beet or mangel crop. Crops sown late, after dispersal has begun, may suffer less injury because the beetle population is smaller and the plants grow faster.

Not all failures occur when beet is grown after a host crop: severe thinning or complete loss of stand sometimes occurs in crops grown in wide rotation, most often in the more intensive beet-growing areas and in fields where the crop grows slowly. However, it is unusual for sufficient beetles to arrive by flight to kill many seedlings. Exceptionally heavy attacks may indicate that the rotational clause has been contravened by overlapping beet with a previous beet or mangel crop.

CONTROL

Rotation is obviously the best way of avoiding damage by this pest. The insecticidal seed dressing used against wireworms gives slight control of pygmy beetle damage. Methiocarb seed dressing is very effective and, after further testing, may be used in the future. Gamma–BHC broadcast and worked into the seedbed gives the best control and is recommended where, under special circumstances and with the British Sugar Corporation's permission, beet follows beet in areas subject to this pest. Since damage fluctuates so much from year to year, depending on the time of migration, preventive treatment with gamma–BHC cannot be recommended generally; in practice it is being used increasingly by growers in the sugar factory areas that suffer most damage from pygmy beetle and millepedes, viz. Peterborough, Ely and Wissington. For crops being damaged, partial control may be obtained by spraying promptly with gamma-BHC at 8 oz of active ingredient (10 fl. oz of 80 per cent suspension) per acre in sufficient water to give some run-off, and preferably concentrated in bands along the rows. Damage to the foliage is rarely severe enough to warrant treatment but may be prevented by spraying with gamma-BHC as above; for this purpose there is less need to concentrate the spray in bands along the rows.

FURTHER READING

BOMBOSCH, VON S. (1963). Untersuchungen zur Lebensweise und Vermehrung von *Atomaria linearis* Steph. (Coleopt. Cryptophagidae) auf landwirtschaftlichen Kulturfeldern. *Z. angew. Ent.* **52,** 313–42.

BONNEMAISON, L. and LYON, J. P. (1968). L'atomaire de la betterave (*Atomaria linearis* Steph.), biologie et méthodes de lutte. *Annls Épiphyt.* **18** (1967), 401–50.

HEIJBROEK, W. (1970). De mogelijkheden voor de bestrijding van de belangrijkste voorjaarsplagen. II. Het bietenkevertje (*Atomaria linearis* Steph.) *Meded. Inst. rat. SuikProd.* **37,** 9–40.

KÜTHE, K. (1971). Neue Möglichkeiten zur Bekämpfung des Moosknopfkäfers (*Atomaria linearis* Steph.). *Zucker* **24,** 219–22.

## SPRINGTAILS

Springtails are primitive insects; they may be globular or elongate, and are usually characterized by a special organ on the underside of the body enabling them to jump a considerable distance. The different species range from white to dark greenish-brown and are about $\frac{1}{20}$ to $\frac{1}{5}$ in. long. Some species feed on the roots and stems of sugar beet, others on the leaves when the weather is warm and moist. The green or yellow leaf-feeders can be mistaken for small aphids, but they are readily distinguished from these by their habit of jumping. They are usually seen on the crop in May or June and their feeding produces minute, rounded pits. The root-feeding springtails (Plate VII) are white; the commonest, *Onychiurus* spp., do not jump when disturbed. When seedlings die because of their attack, the springtails can be found congregated round the main root and stem. Their feeding produces minute, rounded pits which soon darken. Pits on leaves are unimportant but, when numerous on the main root or stem of a tiny seedling, they are dangerous because they provide entry points for pathogenic fungi. The insecticidal seed dressing probably gives adequate protection for all but the most severe infestations; severe damage is rare in England. However, in Belgium, W. Germany and Holland, springtails have recently become a serious pest, possibly because wide seed-spacing and extensive herbicide usage leads to excessive concentration of the creatures on the seedlings.

FURTHER READING

WINNER, C. and SCHÄUFELE, W. R. (1967). Untersuchungen über Schäden an Zuckerrüben durch subterrane Collembolen. *Zucker* **20,** 641–4.

## EARTHWORMS

Earthworms pull seedlings down into their burrows. This happens occasionally in many fields but, where earthworms are particularly numerous, especially in heavy, wet soils and where large amounts of organic matter have been used, damage can be serious. The trouble is most prevalent in wet weather when the earthworms are more prone to come to the surface. No control measures can be recommended.

FURTHER READING

RAW, F. (1961). The agricultural importance of the soil meso-fauna. *Soils Fertil.*, Harpenden, **24,** 1–2.

## Cutworms*

Cutworms or surface caterpillars, which are the larvae of night-flying moths (noctuid or owlet moths), inhabit the surface layers of the soil and feed on many kinds of plants just above or below soil level (Fig. 4(b)). They are dull grey, tinged with various shades of brown, and grow from less than ¼ in. when newly hatched to 1½ in. long when fully grown. Usually they are not seen until they approach 1 in. in length and farmers may confuse them with leatherjackets. Cutworms are typical caterpillars with the usual complement of legs (8 pairs), whereas leatherjackets are legless (Plate I).

### LIFE CYCLE AND PLANT INJURY

The cutworms most injurious to sugar beet are caterpillars of the turnip moth (*Agrotis segetum* (Schiff.)) (Plate I) and the garden dart moth (*Euxoa nigricans* (L.)). The turnip moth is common everywhere; it flies during June and lays eggs in the soil near beet plants towards the end of the month. Eggs hatch at the beginning of July and the tiny cutworms feed on the beet crowns below soil level, causing small round pits, the raw surfaces of which soon darken. As the cutworms grow, the pits become large irregular cavities (Plate IV and Fig. 4(b)). Feeding continues until late autumn. Vigorous crops of sugar beet suffer little injury from this type of feeding, which may be very common in the light soils of East Anglia during long, fine autumns. Backward or late-sown crops may be badly thinned because small tap roots are severed; in such fields, populations of the order of 50,000 cutworms per acre have been observed in mid-July. The cutworms pass the winter in the partly or fully grown state. They feed actively again for a short time in April and May before pupating in May or early June, but the numbers surviving the winter are small enough for this final feeding on sugar beet seedlings to be unimportant to the crop.

The garden dart moth is less common than the turnip moth and crop losses are confined almost exclusively to areas of black fen soil in Huntingdonshire, the Isle of Ely and west Norfolk, where outbreaks of unusual severity sometimes occur. The moths fly during late July and August and lay eggs on the soil of root fields, especially potato fields. Eggs are presumed to remain unhatched until mid-April of the following year, when tiny larvae can be found in early crops of beet, carrots and onions. As a rule, injury to beet passes unnoticed until crops reach the singling stage in mid-May, by which time the cutworms are half grown and begin to destroy seedlings. One cutworm can kill several plants, and as few as 5,000 cutworms per acre are dangerous. This species of cutworm feeds mainly above ground and the tops of seedlings may be completely eaten, leaving only stumps (Fig. 4(b)). Sometimes irregular segments are cut from the cotyledons and early foliage leaves, and the severed pieces may lie beside the damaged plant (Fig. 4(b)). Often all the seedlings in half a yard or more of the row are eaten to ground level with freshly severed pieces, indicating recent feeding, towards one end. The cutworms can often be found by carefully scraping away the soil around a recently injured plant. They usually lie hidden in the soil by day and feed at night, but occasionally feed in the daytime. Feeding continues until mid-June when pupation commences and the cutworms rapidly disappear.

---

* See also Advisory Leaflet 225, available from the Ministry (p. 108).

CONTROL

When cutworms are causing damage, DDT should be applied in the late afternoon or evening, preferably when the soil is moist and the cutworms are likely to be on the surface. Use a dust at not less than 2 lb of active ingredient (e.g., 40 lb of 5 per cent dust) per acre, or a spray at not less than 1 lb of active ingredient (e.g., 3 pints of 25 per cent emulsifiable concentrate) per acre in 20 gal of water. The insecticide should be placed along the rows, if possible, and worked into the soil by close steerage hoeing. Alternatively, 4 oz of DDT (e.g., 1 pint of 25 per cent emulsifiable concentrate), mixed thoroughly with 28 lb of bran and sufficient water to moisten, should be applied as a poison bait at 28–42 lb per acre during the late afternoon or evening; this is convenient for small areas but may be less effective than dusting or spraying. It is more difficult to control turnip moth caterpillars in backward beet in July because they are feeding mainly below ground, but the poison bait has given satisfactory results, indicating that they do wander over the soil. Therefore, DDT dusts or spray applied to the soil around the plants might be effective.

FURTHER READING

PETHERBRIDGE, F. R. and STAPLEY, J. H. (1937). Cutworms as sugar-beet pests, and their control. *J. Minist. Agric. Fish.* **44,** 43–9.

## LEATHERJACKETS*

Leatherjackets, the legless larvae of crane flies or daddy longlegs (Plate I), prefer damp, low-lying grassland. Like wireworms, they cause most injury to crops which follow grass or ley but, because they complete their life cycle in a single season, they cause injury only in the year immediately after ploughing. Also like wireworms, they attack a wide range of crops. They are most prevalent in the wetter districts of Britain and in low-lying marshy land bordering rivers in the drier eastern counties; little beet is grown in these areas, so leatherjackets are of minor importance nationally.

LIFE CYCLE

Of several injurious species of leatherjacket, the commonest are larvae of the marsh crane fly (*Tipula paludosa* Meig.). Adults fly in September and each female lays 200 or more black, slightly elongate eggs in grassland. The eggs hatch after about ten days producing tiny larvae adapted for life in wet soil. Larvae that survive the winter feed actively from March or April until July, when they are from 1½ to 2 in. long. They are then grey and readily distinguishable from cutworms by lack of legs and by their cylindrical shape and rubbery texture. The fully grown larvae pupate in the soil, and the pupae push partly out of the ground just before the flies emerge.

PLANT INJURY

Injury to sugar beet resembles that caused by cutworms or beet carrion beetle, but wounds do not have the ragged, blackened edges caused by the beetle. Irregular segments are eaten from the leaves, growing points may be destroyed and the plants severed just below ground level (Fig. 4(*b*)). Partly grown larvae feed vigorously and can soon kill many plants. Fewer than

---

* See also Advisory Leaflet 179, available from the Ministry (p. 108).

100,000 per acre may ruin a crop and such populations sometimes occur in arable land; populations of a million or more per acre commonly occur in the unploughed marshland of the eastern counties during March and April. Whenever wet, low-lying grassland is to be ploughed for arable cropping, the leatherjacket population should be estimated. This can be done by watering the turf at the rate of 1 gal per sq. yd with DDT-oil emulsion diluted so as to contain 0·025 per cent DDT, which causes the leatherjackets to rise to the surface; an average of 21 per sq. yd is equivalent to 100,000 per acre.

CONTROL

Leatherjackets can be controlled by spraying, dusting or baiting with DDT at the rates recommended for cutworms (page 33), but BHC is preferable. Spray gamma-BHC at 8 oz of active ingredient (10 fl. oz of 80 per cent suspension) per acre, provided that potatoes or carrots are not to be grown within 18 months, and work into the soil during seedbed preparation; or use gamma-BHC in a bran bait at not more than 4 oz of active ingredient (5 fl. oz of 80 per cent suspension) per 28 lb of bran per acre. Paris green at 1 lb, or fenitrothion at $\frac{1}{2}$ lb of active ingredient, per 28 lb of bran per acre can also be used.

FURTHER READING

BARNES, H. F. (1937). Methods of investigating the bionomics of the common crane-fly, *Tipula paludosa* Meigen, together with some results. *Ann. appl. Biol.* **24,** 356–68.

## SLUGS*

Slugs, especially the field slug (*Agriolimax reticulatus* (Müll.)), feed on sugar beet plants at any time of year and can readily kill beet seedlings and seriously damage small stecklings. Slugs are imperfectly adapted to land life and can be active only when the soil or the air is moist. The field slug usually feeds on foliage in the late evening and at night, but sometimes also during the day in damp, dull weather; at other times it must shelter where moist, i.e., in the soil or under vegetable litter. Slugs are usually more numerous where drainage is imperfect and where the land is heavy and cloddy; they are favoured by the use of organic manures, by crops that provide ample shelter and cover, by the presence of crop debris, and by moist, mild conditions. Numbers in the soil may well reach 100,000 per acre in the spring when sugar beet is in the seedling stage and easily damaged; even 5,000 per acre can severely thin a widely spaced stand. Damage was particularly prevalent in many beet-growing areas in May, 1966. Generally, damage is noticed only in some heavy-land areas of England.

PLANT INJURY

Damage to individual plants is not particularly characteristic and can be confused with that caused by cutworms, leatherjackets or sand weevils (Figs. 4(a) and (b)). The field slug eats irregular areas from the leaves and stem (Plate III) and, in the course of a night or two, only the root remains. Slime trails are only rarely visible and the slugs are very difficult to find in the soil. The white-soled slug (*Arion fasciatus* (Nilss.)), another species known to damage beet, feeds below soil level and severs the root; the plant wilts and

---

* See also Advisory Leaflet 115, available from the Ministry (p. 108).

dies but the culprit is not always easily found. If slug damage is suspected, search the area with a torch at nightfall for surface-feeding slugs. Alternatively, or in addition, a bait containing metaldehyde or methiocarb (see below) can be scattered on several small areas, each of a few square yards, and inspected the following morning. Slugs affected by these baits remain on the soil surface and can readily be seen. An average of one or more per square yard indicates that treatment is needed promptly. In fields where damage can be expected, such a check is best made soon after drilling, provided that conditions favour slug movement on the soil surface; treatment should be applied where necessary to avoid the probability of subsequent loss of plant.

CONTROL

For sugar beet, poison baits are preferable to sprays. Use $\frac{1}{2}$ lb of metaldehyde to 28 lb of bran or beet pulp per acre and place in small discrete heaps by hand. Proprietary baits in pelleted form are very convenient; they should be spread by hand or machine, e.g., fertilizer distributor of the 'plate and flicker' or 'star wheel' type, at the rates recommended by the manufacturer, usually 28 lb per acre. Very good results with rates as low as 10 lb per acre have been achieved by using a cup-feed drill to place one row of pellets along each row of beet. Methiocarb as a proprietary bait at only 5 lb per acre is equally, or even more, effective. In badly infested fields, slugs in all stages of growth as well as eggs are present; treatment will save the crop but will not prevent numbers increasing again later.

FURTHER READING

BARNES, H. F. and WEIL, J. W. (1944, 1945). Slugs in gardens: their numbers, activities and distribution (Parts 1 and 2). *J. Anim. Ecol.* **13,** 140–75 and **14,** 71–105. Part 1 contains a key for the identification of our common species and part 2 has coloured illustrations of them.

GOULD, H. J. (1962). Trials on the control of slugs on arable fields in autumn. *Pl. Path.* **11,** 125–30.

JONES, F. G. W. and JONES, M. G. (1964). *Pests of field crops.* London, Edward Arnold.

WEBLEY, D. (1969). A comparison of methiocarb and metaldehyde baits for the control of four species of slugs. *Proc. 5th Br. Insecticide & Fungicide Conf. Brighton, 1969,* **2,** 442–4.

## CHAFER GRUBS*

Chafer grubs or white grubs, the larvae of chafer beetles, are damaging pests in a few well-wooded areas where the soil is light, e.g., the Breckland areas of Norfolk and Suffolk. The species that most often attacks sugar beet is the cockchafer (*Melolontha melolontha* (L.)) but occasionally the summer chafer (*Amphimallon solstitialis* (L.)) is damaging. Attacks by the garden chafer (*Phyllopertha horticola* (L.)) are rare. Chafer grubs (Plate I) are curved, white and fleshy, and are $\frac{1}{10}$ to $1\frac{1}{2}$ in. or more long.

LIFE CYCLE

The life cycle of the cockchafer occupies three years. Adult beetles emerge from the soil in May and June and swarm in trees, flying actively at dusk.

---

* See also Advisory Leaflet 235, available from the Ministry (p. 108.)

They lay eggs beneath the soil, usually in grassland or cereals but also in arable land. These eggs produce small larvae that become fully grown in June–August of the third year. They pupate in the soil and the pupae develop into adults later that year. The adults overwinter in the soil until May or June of the following year, thus completing the cycle. Sugar beet is damaged by the larger larvae, especially those in their second and third year, but larvae in all stages can usually be found where damage is occurring. Damage is more frequent after grass or cereals than other arable crops. The life cycle of the summer chafer is similar to that of the cockchafer but occupies only two years.

PLANT INJURY

Chafer grubs feed entirely below ground at a depth of 2–12 in. Out of the soil they seem cumbersome and helpless, but in it they burrow actively and use their powerful jaws to feed on the roots of various plants. Injury to sugar beet is first noticed shortly after singling and continues through the summer into the autumn. Attacked plants usually wilt and may die, and their roots have extensive wounds that turn black, with the insect's jaw marks visible on the raw surfaces (Plate VI and Fig. 4(b)). Large plants suffer less from these wounds, but fully grown beets may be killed when the tap root is severed some 8–12 in. below ground. Adult chafer beetles feed mainly on the foliage of trees, but garden chafer adults occasionally swarm in sugar beet during June and feed on the leaves.

CONTROL

Because outbreaks are very local and sporadic, control measures are difficult to devise. Beet should preferably not be grown in fields where many chafer grubs are seen during ploughing and cultivation, because even a few may damage crops seriously; rotavation should kill many grubs. In France and Switzerland cockchafers are a serious problem; large tracts of woodland have sometimes been sprayed with DDT from the air to kill the adults while they are feeding in May and June before the eggs are laid. Because the life cycle occupies three years there are three independent 'streams', of which the adults fly in different years. In some districts one 'stream' is dominant, in others another, so that mass flights of adults occur in different years. This phenomenon has not been observed in Britain. The persistent organochlorine compounds cannot be used for control because of the large amounts that would be needed, the difficulties of incorporation in the soil, and the residues that would be left. Granular organophosphorus insecticides placed near the growing plants might control chafer grubs, but have not been tried.

FURTHER READING

FIDLER, J. H. (1936). Some notes on the biology and economics of some British chafers. *Ann. appl. Biol.* **23**, 409–27.

KELLER, E. (1954). Erfahrungen mit der chemischen Maikäferbekämpfung in der Schweiz. *Anz. Schädlingsk.* **27**, 147–52.

MILNE, A. (1956). Biology and ecology of the garden chafer, *Phyllopertha horticola* (L.). II. The cycle from egg to adult in the field. *Bull. ent. Res.* **47**, 23–42.

## Swift Moth Caterpillars*

Swift moth caterpillars (*Hepialus* spp.) inhabit soil and feed entirely below ground like wireworms and chafer grubs. When fully grown they are about 1¼ in. long with a reddish-brown, hard head and a translucent white, fleshy body which is prominently segmented and bears only a few, widely scattered hairs (Plate I). The moths are moderately large with light brown or greyish wings flecked with a white pattern. They fly in the evening and prefer to lay their eggs in clumps of perennial plants such as strawberries, pyrethrums, Michaelmas daisies and weeds. Consequently, caterpillar injury to crop plants is more common in gardens and market gardens than in the open field, but sometimes cocurs where land has been allowed to become foul with weeds in the previous year. The caterpillars feed on many plants, wounding or severing roots and stems below ground. They may, like cutworms, sever the hypocotyls of seedling sugar beet or make pits in the crowns of older plants, but they never feed above soil level. Like chafer grubs, they are difficult to combat with insecticides, but fortunately they are very rarely sufficiently numerous on sugar beet to warrant control measures.

FURTHER READING

EDWARDS, C. A. and DENNIS, E. B. (1960). Observations on the biology and control of the garden swift moth. *Pl. Path.* **9,** 95–9.

## Rosy Rustic Moth Caterpillar

This caterpillar, also known as the potato stem borer (Plate II), is the larva of a dingy, night-flying moth known as the rosy rustic moth (*Hydraecia micacea* (Esp.)), which is similar in many respects to the turnip moth (Plate I). The moths fly in late summer and autumn and lay their eggs on bare soil at this time. The fleshy caterpillars are dull pink with dark brown raised spots along both sides; from each spot a fine hair grows. The caterpillars feed by burrowing into the rootstocks and stems of many plants including sugar beet, potato, and weeds such as dock and plantain. They sometimes burrow in young beet plants in late May, June and early July, causing wilting and collapse (Plate V and Fig. 4(*b*)). Attacks are sporadic and only isolated plants are killed; such plants are normally on field headlands. No control measures are possible or necessary.

FURTHER READING

SHAW, M. W. (1957). Damage by rosy rustic moth larvae in Scotland, 1956. *Pl. Path.* **6,** 135–6.

# LEAF-EATING PESTS

## Beet Flea Beetle†

The beet or mangold flea beetle (*Chaetocnema concinna* (Marsh.)) is distinct from cruciferous or turnip flea beetles (*Phyllotreta* spp.) which attack turnips, swedes, kale and other cruciferous crops, although it is often confused with them. Cruciferous flea beetles do not attack sugar beet but can often be found

---

\* See also Advisory Leaflet 160, available from the Ministry (p. 108).

† See also Advisory Leaflet 109, available from the Ministry (p. 108).

in healthy beet crops feeding upon cruciferous weeds such as charlock. Beet flea beetle occurs in all beet-growing areas. In some years damage used to be widespread and many acres had to be resown. On the whole, outbreaks were more common on the drier, eastern side of the country. Severe injury has been rare since 1962, perhaps because of the dieldrin seed dressing, but moderate injury was widespread in spring 1971.

The adult flea beetles are about $\frac{1}{10}$ in. long and bronze in colour with rows of deep punctures on the wing cases (Plate II). Like other flea beetles, their hind legs have enlarged femora which project on either side of the body and enable the beetles to jump when disturbed. They feed on crops and weeds in the dock family (*Polygonaceae*), e.g., rhubarb, buckwheat, knotgrass (*Polygonum aviculare* agg.), black bindweed (*Polygonum convolvulus* L.) and persicaria (*Polygonum persicaria* L.), as well as on sugar beet and mangel.

LIFE CYCLE

The beetles hibernate in tufts of grass, overgrown hedgerows, shelter belts, woods and other deep cover. On the southern side of woods and shelter belts, beetles may emerge in warm weather at the end of March or in early April to feed on early-sown crops and weeds; widespread dispersal occurs only with the first spell of warm, sunny weather when air temperatures approach 70°F (21°C) and wind speeds are below about four miles per hour. Then the beetles start feeding on seedling beet in fields remote from their winter shelter; this is usually in mid-April and feeding continues into May and June. Beetles feed extremely vigorously on warm, clear, sunny days when the soil surface dries rapidly and surface temperatures reach high levels. At such times serious damage may occur, especially if the deeper soil is cold and growth is slow. Seedlings in the cotyledon stage may be crippled in two or three days if beetles are numerous. Eggs are laid in the soil near the host plants and hatch into tiny white larvae which feed on the seedling roots without causing apparent injury. The larvae live for about six weeks and then pupate in the soil. Numerous new beetles emerge from July onwards and feed on their host plants, but do not seriously injure them because the beet and mangels are well grown by this time.

PLANT INJURY

Flea beetle injury is fairly characteristic (Plate IV and Fig. 4(a)). The beetles feed on either surface of the leaves, producing circular pits that do not penetrate the far surface, which is thus left intact but may fall away later as the leaf grows. When feeding is excessive, the pits coalesce, the plant is defoliated and the growing point may be destroyed. Young seedlings are very susceptible but older seedlings are able to withstand attack, especially if they are growing rapidly.

CONTROL

Flea beetles may be controlled readily with DDT but, if maximum benefit is to be obtained, this must be applied as soon as the first signs of injury appear. In areas where the pest is prevalent, it is advisable to keep a close watch on seedling crops, especially in periods of fine weather. The most serious damage is caused while the seedlings are in the cotyledon stage. DDT should be used at 8–16 oz of active ingredient (e.g., 2–4 pints of 20 per cent emulsifiable concentrate or 25–50 lb of 2 per cent dust) per acre, depending on the severity of attack. Even 5 gal of spray per acre are adequate.

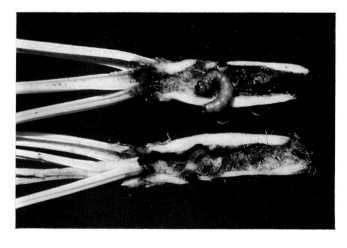

A young beet plant with the root tunnelled by the potato stem borer, the caterpillar of the rosy rustic moth.

Foliage injury due to feeding by beet carrion beetle. Two nearly full-grown larvae can be seen on the leaves. Note the blackened, irregular edges of the areas cut from the leaves.

Blister mines in leaves of sugar beet caused by the tunnelling of mangold fly larvae (beet leaf miners) between the upper and lower surfaces.

PLATE V

Severe foliage feeding by the pygmy beetle. This sometimes occurs in the seedling during spells of moist weather. In dry weather feeding occurs below ground only. Some wounds are symmetrically placed about the midrib: feeding occurs while the leaf is still folded.

Partly grown beets with their taproots severed by the feeding of a chafer grub. Feeding often extends up the root as on the left-hand plant.

Capsid injury to sugar beet. The terminal portion of the leaf blade has yellowed following stab wounds on the main vein just below the yellowed area.

PLATE VI

Black shiny eggs of the black bean aphid laid about the buds of spindle tree.

Immature black bean aphid feeding on plant. (From *Pests of Field Crops* by F. G. W. Jones and M. G. Jones, 1964.)

Spotted snake millepede (x 6).

Glasshouse symphylid, *Scutigerella immaculata* (x 7).

Springtails (x 20). These are commonly found feeding on the roots of seedlings and are often associated with symphylids and mille-pedes.

PLATE VII

Galls on lateral roots caused by a root-knot eelworm, *Meloidogyne naasi* (x 1½).

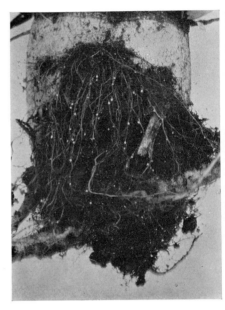

Plant stunted and bearded by heavy beet cyst eelworm attack. Note the numerous white female eelworms which will become brown cysts (x ½).

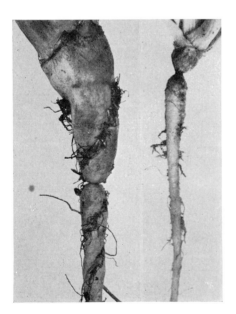

Strangles. Right: a plant showing the typical constriction at soil level due to drying at an earlier stage. Left: a plant showing excessive 'corkscrewing' of the root and 'deep strangles', possibly due to some chemical toxicity.

Female cyst eelworms protruding from the fibrous roots (x 25).

PLATE VIII

DDT persists after application, so that a second application is rarely necessary, except when heavy rain falls immediately after treatment. Seed dressings containing much gamma-BHC are used for the control of flea beetles on cruciferous crops. On no account whatever should these be used on sugar beet; the plants may be poisoned, with serious consequence to the crop.

FURTHER READING

NEWTON, H. C. F. (1929). Observations on the biology of some flea-beetles of economic importance. *Jl S.-east agric. Coll. Wye* **26,** 145–64.

STAPLEY, J. H. (1949). *Pests of farm crops.* London, Spon.

## SAND WEEVIL

The sand weevil (*Philopedon plagiatum* (Schall.)) is widespread in England and Scotland, especially near the sea, but has been reported as a pest of sugar beet only in the Breckland areas of west Norfolk and west Suffolk and in one or two separate localities around the East Anglian coast. It is also a pest of beet in Europe. The adult insect (Plate II), which is unable to fly, is about $\frac{1}{4}$ in. long, and the wing cases are thickly covered with brown scales that are alternately light and dark, giving an appearance of longitudinal stripes.

LIFE CYCLE AND PLANT INJURY

The weevils overwinter in the soil from which they emerge in April. They begin to feed at once, continue until early June and, by mid-June, most have disappeared. At the end of May and beginning of June, they lay eggs in the soil around beet plants and weeds. White, legless larvae hatch and feed upon the roots. These larvae live for about 18 months, eventually developing into pupae in the soil, from which the adults emerge in the following year.

Adult weevils feed on many kinds of plants. When they feed on sugar beet, which is by no means their favourite host, they cut semicircular areas from the edges of the cotyledons and leaves (Fig. 4(*a*)). They feed mainly at night, but also on dull moist days, and generally near the soil surface. Consequently, most damage on well-grown plants is near the leaf stalks or near the leaf tips where these bend back and touch the ground. Severe injury to seedling beet is attributable to the following factors: (1) the coincidence of the intense feeding period of the weevils with the susceptible seedling stage; (2) the wide spacing and relatively small plant population; (3) the absence or relative scarcity of weeds in the crop, especially where herbicides are used; and (4) the position of sugar beet in the rotation in areas where the sand weevil occurs. Injury to beet may be expected when it is preceded by three consecutive crops drawn from the following two groups:

(*a*) Cereals: wheat, oats, barley;

(*b*) Leys: sainfoin, lucerne, clover, ryegrass.

The following rotation on one farm regularly gave rise to attacks:

barley (undersown); sainfoin; winter wheat; sugar beet.

Populations of 5,000 weevils per acre in April are troublesome; populations of 50,000 or more per acre cause total failure.

CONTROL

The pest is readily controlled by DDT used at the rates recommended for beet flea beetle (see page 38); prompt treatment is essential.

## Beet Leaf Weevil

The beet leaf weevil (*Tanymecus palliatus* (F.)) (Plate II) is closely allied to pea and bean weevils (*Sitona* spp.) but feeds on beet and its allies as well as on leguminous crops. Like the pea and bean weevils, it is a leaf-edge feeder and cuts semicircular segments from the cotyledons and early foliage leaves, working from the margins inwards. In dry seasons sporadic injury occurs in many places in Europe, especially in Hungary and Rumania, and when beet follows lucerne. Some unconfirmed reports of pea and bean weevils feeding on sugar beet in England may, in fact, be this related weevil; it has been caught in pitfall traps in English beet fields.

## Beet Carrion Beetle

The beet carrion beetle (*Aclypea opaca* (L.)) is widespread on the Continent and in Britain, where it was first noted in 1888 on mangels. Occasionally, when the pest was abnormally abundant, attacks occurred in most beet-growing areas, but no serious ones have been reported for more than eighteen years. Carrion beetles are so called because they feed on decaying animal matter. The beet carrion beetle, however, prefers a vegetable diet and attacks many other plants in addition to beet and mangels. Plants in the beet family (*Chenopodiaceae*) are most favoured and it is only on them that the insect is a pest. Other species of carrion beetle attack beet on the Continent, but records of this happening in Britain are few and require confirmation.

LIFE CYCLE AND PLANT INJURY

The adult beetles are about $\frac{1}{2}$ in. long and generally dull black (Plate II) but occasionally bronze-coloured. They spend the winter in the shelter of fallen leaves, dry grass, moss, brushwood and other litter, especially favouring the borders of woods and copses. Beetles emerge from hibernation when the weather is fine in mid-April and move into fields where they start to feed on crops. Early-sown beet sometimes suffers in the seedling stage from adult beetles feeding on the cotyledons. They lay white, spherical eggs in the soil from the end of April onwards, and black, obviously segmented larvae that resemble woodlice hatch from them in a few days. The larvae feed freely on the leaves of seedling beet and mangels through May, June and early July. They become fully grown in two to three weeks and give rise to white pupae in the soil from which the next generation of beetles emerges after ten to fourteen days. There is one generation in the year but, because of the protracted period of egg laying, adults and larvae in all stages can usually be found feeding together. They feed most vigorously in warm, dry weather and, when disturbed, leave the plant hurriedly to hide under fallen leaves, stones, or in cracks in the soil. Farmyard manure appears to attract the adult beetles and sometimes attacks are limited to the manured portions of fields, but severe attacks occur as frequently on manured as on unmanured land.

Injury consists of irregular holes and segments, with characteristically blackened raw edges, cut in the leaves of beet and mangels (Plate V and Fig. 4(c)). When the insects are numerous, the plants are defoliated and may be killed.

CONTROL

The pest is so rare that control has been unnecessary for very many years. Should there be a recurrence of damage sufficient to warrant control, use gamma-BHC at 8 oz of active ingredient (e.g., 10 fl. oz of 80 per cent suspension) per acre.

FURTHER READING

JONES, D. P., PETHERBRIDGE, F. R. and JENKINS, A. C. (1947). Beet carrion beetle in England and Wales. *Agriculture, Lond.* **54**, 375–7.

## BEET LEAF MINER*

Beet leaf miner, the larva of the mangold fly or beet fly (*Pegomya betae* (Curt.)), is present in most years in certain districts, notably in the coastal regions of East Anglia and in some parts of Lincolnshire, Nottinghamshire, Staffordshire, Shropshire and Yorkshire. Occasionally attacks are severe and widespread; the reasons for variations in incidence are probably climatic. German scientists have shown that the fly is favoured by cool, moist weather, and in places where summers are hot and dry it is not a serious pest. The regions mentioned above are all somewhat cooler and moister in summer than the inland parts of East Anglia and the south. They are also, for the most part, regions of light sandy soil, which may also have some influence on the occurrence of the pest.

LIFE CYCLE

This insect overwinters in the soil in the pupal stage. The adults (Plate II), which are light grey and similar in appearance to the house fly, usually emerge from the soil during the latter half of April, the time differing a week or so with the season and the district (they are later in the north than in the south); soil temperature is probably the determinant. The flies emerge from the old beet fields and disperse. The first eggs are laid in new beet fields at the end of April and early in May. The eggs are deposited singly or in batches of from two to twenty on the undersides of the cotyledons and young leaves. In captivity a female may lay as many as 300 eggs, but in the field fewer are probably laid. Egg laying continues for a month or more, and in East Anglia the greatest number of eggs per plant is reached about the end of May.

After four or five days tiny legless larvae (maggots) hatch from the eggs, burrow straight into the leaf and mine between the upper and lower surfaces causing characteristic blisters. If the leaves are small or mining is extensive, the larvae may wander over the surface to fresh leaves. When fully grown at the end of ten to fourteen days, the larvae drop to the soil, burrow below the surface and turn into brown, barrel-shaped pupae. The second generation of flies emerges from these pupae after two to three weeks, producing a second series of mines on beet leaves in July. In most years a partial third generation of flies causes mines in August-September. The second and third generations rarely injure the crop seriously, but they may produce many pupae which overwinter in the soil and start an outbreak in the following year. Often, however, the second and third generations are decimated by adverse weather or parasites killing the larvae or pupae.

PLANT INJURY

The injury caused by the larvae mining in the leaves is characteristic (Plate V and Fig. 4(*c*)). At first, pale, winding, linear mines are produced,

---

* See also Advisory Leaflet 91, available from the Ministry (p. 108).

which the larvae quickly enlarge into characteristic blisters within which they may be found. The blisters eventually wither so that the leaves have a brown, scorched appearance. If the seedlings are small and the attack severe, the whole leaf surface may be destroyed. However, sugar beet will tolerate considerable defoliation and, provided the terminal bud is intact, will suffer only a temporary check. Severe attack by leaf miner on well-grown plants may decrease the yield by one ton per acre of washed beet; on backward plants, losses may exceed two to three tons per acre (see page 4 for effects of artificial defoliation).

CONTROL

Yield loss is best avoided by early drilling on soil adequately manured and in good tilth; this provides plants that are large and vigorous at the time when the first generation of this pest attacks. If the crops are singled before or during egg laying, the flies concentrate the eggs on the remaining plants; if singling is after the egg-laying period, many eggs and larvae are destroyed with the plants that are chopped out, and the plants selected to stand have fewer eggs on them. The wider seed spacings now being adopted by growers inevitably lead to a greater concentration of eggs on fewer plants, and sowing to a stand (30,000–40,000 seedlings per acre) increases the likelihood of more eggs and larvae per seedling and hence the risk of yield loss due to this pest.

The easiest stage to control with insecticides is the larva mining in the leaves; action should be taken when egg laying seems complete and when the first mines are appearing. None of the insecticides recommended below kills the eggs; they kill only the larvae as they hatch from the eggs or mine inside the leaves. The level of attack at which it pays to spray is difficult to determine, because it depends on the size and rate of growth of the plants. However, when the number of new, unhatched eggs plus living larvae equals or exceeds the square of the number of rough leaves, the pest is likely to diminish yield. It can be controlled by spraying, preferably at 20 or more gallons per acre, with one of the following materials: dimethoate at 1·2 oz of active ingredient (e.g., 3 fl. oz of 40 per cent emulsifiable concentrate) or formothion at 1·9 oz of active ingredient (4 fl. oz of 43 per cent emulsifiable concentrate) or trichlorphon at 6·4 oz of active ingredient (8 oz of 80 per cent soluble powder) per acre. These all kill rapidly and are suitable for any level of attack or stage of larvae. Only top cover is required as the insecticides, whether systemic or not, penetrate the leaves locally and kill newly hatched larvae for some days after spraying.

Probably no crop past the eight-leaf stage is worth spraying to control this pest alone, but in years when aphids invade early the two pests can be simultaneously checked by using dimethoate at 4·8 oz of active ingredient (e.g., 12 fl. oz of 40 per cent emulsifiable concentrate) per acre, which controls leaf miner excellently and green aphid well, although an increased dosage is necessary to control the black bean aphid; alternatively, formothion at 7·5 oz of active ingredient (16 fl. oz of 43 per cent emulsifiable concentrate) or phosphamidon at 3·2 oz of active ingredient (e.g., 16 fl. oz of 20 per cent emulsifiable concentrate) per acre may be used. A mixture of 3·2 oz of trichlorphon and 3·4 oz of demeton-S-methyl (4 oz of 80 per cent soluble powder plus 6 fl. oz of 58 per cent emulsifiable concentrate, respectively) per acre controls both these pests and the yellows viruses well. Trichlorphon should not be used alone when any aphids are present as it kills aphid enemies better than aphids.

FURTHER READING

DUNNING, R. A. (1961). Mangold fly incidence, economic importance and control. *Pl. Path.* **10,** 1–9.

DUNNING, R. A. and WINDER, G. H. (1965). The effect of insecticide applications to the sugar beet crop early in the season on aphid and yellows incidence. *Pl. Path.* **14,** Suppl. 30–6.

JONES, F. G. W. and DUNNING, R. A. (1954). The control of mangold fly (*Pegomyia betae* Curtis) with DDT and other chlorinated hydrocarbons. *Ann. appl. Biol.* **41,** 132–54.

WINDER, G. H. (1971). Control of beet leaf miner (*Pegomya betae* (Curt.)) by soil-applied pesticides. *Pl. Path.* **20,** 164–6.

## TORTRIX MOTH CATERPILLARS

Caterpillars of various species of tortrix moth feed on beet foliage in May and June of most years, but only in some years are numbers sufficient for damage to be readily noticeable. They are small and conceal themselves by binding parts of a leaf or leaves together, confining their feeding to the leaf surfaces within (Fig. 4(*c*)). They move rapidly when disturbed. The caterpillars of the commonest species, *Cnephasia interjectana* (Haw.), are about $\frac{3}{8}$ in. long when fully grown, greyish-green with small black spots, and taper slightly to a dark head and tail. The mature caterpillars pupate where they have been feeding and the small greyish-brown moths emerge about three weeks later.

Damage is more apparent than real and control measures are rarely necessary. Trichlorphon, as used for beet leaf miner, at not less than 6·4 oz of active ingredient (8 oz of 80 per cent soluble powder) per acre kills the caterpillars readily.

## DIAMOND-BACK MOTH CATERPILLAR

The caterpillar of the diamond-back moth, *Plutella xylostella* (L.), is a fairly common pest of cruciferous crops. It damages beet regularly in Russia but only in certain years in this country. One such year was 1958 when vast numbers of moths arrived at the end of June, presumably from Russia or Scandinavia. In July the moths were found readily in beet crops in eight factory areas, almost exclusively those on the east coast, but the caterpillars did little damage because leaf growth was rapid.

The moth is thin, about $\frac{3}{8}$ in. long and light brown with three characteristic yellowish diamond-shaped patches along its back; it is easily seen in the crop when disturbed. The caterpillar is pale green with a darker head; it tapers towards each end and is $\frac{1}{2}$ in. long when fully grown. It feeds on the underside of the leaf, tunnelling when small but feeding on the surface later; when disturbed it wriggles off the leaf and suspends itself by a silken thread, later returning to the leaf. The caterpillar pupates in a fragile, silken cocoon on the underside of the leaf.

Damage caused is negligible and control measures are unnecessary.

FURTHER READING

FRENCH, R. A. and WHITE, J. H. (1960). The diamond-back moth outbreak of 1958. *Pl. Path.* **9**, 77–84.

## SILVER Y MOTH CATERPILLAR

The silver Y moth (*Plusia gamma* (L.)) belongs to the same group as the turnip moth and the garden dart moth; unlike these, however, it flies by day. It may readily be distinguished by the prominent silver Y marking on each fore-wing from which it derives its name (Plate I). The moths migrate, travelling many hundreds of miles; the first to arrive in this country probably come from the Mediterranean, but many more cross the North Sea later, in July and August, which may explain the severity of attacks on beet in eastern coastal areas. Moths first appear in late May and may be seen sipping nectar from flowers. They lay eggs in June and these develop into tiny caterpillars. The holes caused by their feeding become conspicuous in late July and early August (Plate III and Fig. 4(c)). The caterpillar is light green and possesses only three pairs of additional legs on the abdomen instead of the usual complement of five extra pairs. It feeds on many plants including sugar beet and, when mature, spins a loose silken cocoon on the foliage where it has been feeding and pupates within.

Widespread infestations occur in some years, e.g., 1936 and 1946. In such years beet leaves are skeletonized in late July and August in many districts. Unless damage is very severe—loss of at least one third of the leaf area—recovery is rapid and control measures are unnecessary. The caterpillars disappear rapidly towards the end of August, when some pupate and many are killed by parasites or disease. All stages die out in the late autumn for the species seems unable to overwinter in Britain, even in the milder south-west.

## EARWIG

The common earwig (*Forficula auricularia* L.) injures garden plants such as dwarf beans, cucumbers, cauliflowers, tomatoes and many cultivated flowers. Field crops are less often attacked, but leaves of potato and especially sugar beet are injured. The insect is unusual in that the female tends the eggs and young. Eggs are laid in autumn and spring in a nest or pocket in the soil. They hatch in April and the young stay in the nest for some time, gradually foraging further afield. Later the family breaks up but, nevertheless, earwigs tend to remain in large groups. Feeding occurs by night but earwigs can be found hiding at the bases of attacked plants or under clods of earth nearby. They eat irregular holes in the heart leaves of sugar beet in June or early July (Fig. 4(c)). Sometimes injury appears severe on fields adjacent to waste land, scrub, or railway embankments. Control measures are unnecessary.

FURTHER READING

BURR, M. (1939). Modern work on earwigs. *Sci. Prog., Lond.* **34**, 20–30.

## TORTOISE BEETLES

Tortoise beetles are so called because of their shape (Plate II). Three species occur on sugar beet in Britain: *Cassida nobilis* L., *Cassida vittata* de Vill. and *Cassida nebulosa* L., of which the last is the least common. *Cassida* spp. are sometimes serious pests in Italy, Spain and Turkey but are unimportant in Britain. They hibernate in shelter and then reappear in beet fields in April where they feed upon beet seedlings and related weeds, cutting segments from the cotyledons and shot-holing the young leaves. Eggs are laid from May onwards and the larvae which hatch are pale green, possess forked tails and bear lateral spines on each segment. After moulting, the larvae carry the cast skins, which adhere to the tail, over their backs. Small larvae cause injury which looks similar to that made by flea beetles: circular pits are eaten in the lower surface of the leaf but the upper surface is left intact. When the beetles or their larvae are numerous, they may devour the leaves except for the veins, but such severe injury is rare. *C. nebulosa* and *C. vittata* may produce two generations per year but *C. nobilis* only one.

## SAP-SUCKING PESTS

Sap-sucking pests pierce the plant tissues with their needle-like mouth-parts, inject saliva and extract sap. This group includes aphids (greenflies, blackflies, etc.), capsids (mirids) and leaf bugs, as well as thrips and red spider mite, which are small creatures whose sucking mouthparts do not penetrate deeply. All cause mechanical injury to the plant tissues they pierce and weaken the plant by extraction of sap; sometimes their saliva kills tissue around the wound or induces leaf curl. However, the most serious injury occurs indirectly when a sucking insect transmits a virus which spreads throughout the plant and impairs its growth. Whereas many small insects are needed to injure a plant seriously by feeding, a single one may infect the whole plant with a virus.

Aphids are probably the most mobile of sugar beet pests, especially as on suitable days those in flight are carried passively on up-currents of air and later settle many miles from where they took to wing.

The types of injury caused by the sap-sucking pests are:

Black bean aphid —some virus transmission but mainly direct injury by sap-sucking, cell destruction and leaf curl on young and old plants.

Peach-potato aphid —some direct injury but mainly virus transmission.

Other aphids —minor direct injury and minor virus transmission.

Capsids —direct injury due to cell damage and toxic saliva, death of growing points of young seedlings, and distortion and yellowing of leaf extremities of plants in summer.

Thrips
Leafhoppers ⎫
Red spider mite ⎬ superficial injury to leaf surface.
⎭

### BLACK BEAN APHID*

Severe and widespread epidemics of the black bean aphid (*Aphis fabae* Scop.) occur in some years (Figs. 7(a) and (b)). If uncontrolled on seed crops, the

---

* See also Advisory Leaflet 54, available from the Ministry (p. 108).

aphid can limit seed production because heavily infested crops give poor yields of seed. Attack on the root crop is less devastating but, nevertheless, greatly decreases yield. It is difficult to assess losses as epidemics usually coincide with prolonged drought and virus yellows. Losses caused by the feeding of the aphids have been estimated at two to four tons of beet per acre; when they spread beet yellows virus, the losses are much greater.

PLANT INJURY

Physical damage to tissue of the lower surface of the leaf, loss of sap and injection of toxic saliva curls the leaves severely and causes wilting and stunting of the plants. As the aphids multiply, leaf curling increases so that the aphids are sheltered, thus adding to the difficulty of killing them with insecticide. Colonies in the hearts of sugar beet so injure the young leaves that they never reach their normal size and parts of the leaves wither giving the plant a scorched appearance. Damage to young heart leaves is always more severe than to the older, fully expanded, outer leaves. A sticky fluid known as honey-dew, which is excreted from the gut of the aphids, coats the leaves and encourages growth of sooty moulds which prevent photosynthesis. Black aphids may cause severe losses of yield if they infect the plants with yellowing and beet mosaic viruses, but they spread these viruses less efficiently than the peach-potato aphid and do not transmit beet mild yellowing virus.

DESCRIPTION AND LIFE CYCLE

The black bean aphid is easily distinguishable from the other aphids found on sugar beet (Plate VII and Figs. 10(a) and (b)). It is entirely black, except for patches of white wax on the abdomen of some young. The siphunculi or cornicles (the pair of tubes found on the upper surface towards the rear of the abdomen and characteristic of aphids) are short and straight.

The life cycle of this aphid, like that of many others, is complicated (Fig. 6). It passes the winter as small, black, shiny eggs on the spindle tree (Euonymus europaeus L.) (Plate VII) and on the sterile guelder rose (Viburnum opulus roseum L.) grown in gardens. Young wingless females hatch from the eggs in March and early April, when the bushes are bursting into leaf, and feed on the young shoots and developing leaves. In about three weeks these females, without previous fertilization, begin to produce living young, i.e., parthenogenetically and viviparously. The females of the new generation also reproduce in the same way and soon colonies are formed which curl the spindle tree leaves in a characteristic manner. At the end of April or during May, according to the season, winged females develop in the colonies. After this the aphids produced are mostly winged and the spindle tree may be free of aphids by the first week in June or even earlier. The winged aphids, known as *spring migrants*, fly to the various summer hosts such as beans, sugar beet and mangels. The spring migration usually starts in early May and continues into June, reaching a peak in mid-May. The terminal buds of beans and the flowering shoots of sugar beet, mangel and red beet seed plants are especially susceptible to infestation. On their summer hosts, the winged migrants produce living young that are wingless. These young remain on the host plant and soon large colonies are formed and winged aphids appear once more. These winged aphids are the *summer migrants* and they continue the colonization of summer hosts begun by the spring migrants.

Throughout the summer the aphids give birth to wingless young, but in July colonies are usually much decreased by aphid enemies such as ladybird beetles and their larvae, hover fly larvae, lacewing larvae, certain

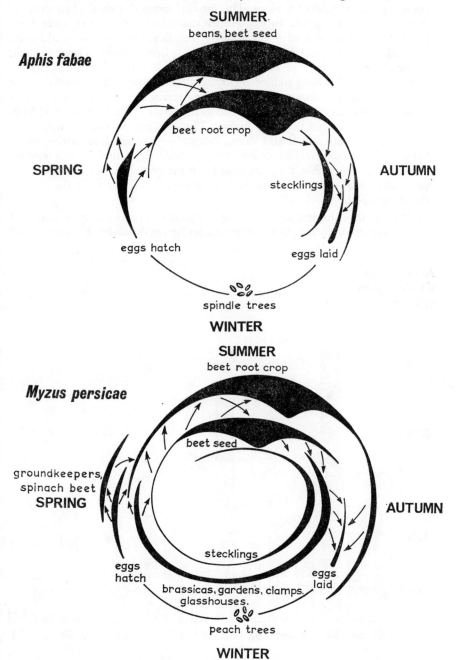

Fig. 6. Life cycles of the black bean aphid (upper) and the peach-potato aphid (lower). The thickness of the lines indicates approximately the differences in abundance and the arrows indicate migration between crops

parasitic wasps, and entomogenous fungi. Consequently, the aphid population declines and, the greater the infestation, the more rapid the decline;
live aphids may be difficult to find only two to three weeks after there have
been as many as 5,000 per plant. This usually occurs during the last weeks of
July or early in August. In September the colonies on the summer hosts
begin to produce winged *return migrants* and, later, winged *males*, although
some black aphids can often be found on sugar beet as late as December and
a few may overwinter in mangel clamps. The return migrants fly to the
spindle tree and there produce *egg-laying females*. The males also fly to the
spindle tree where they mate with the egg-laying females; this is the only
occasion when sexual reproduction occurs in the life cycle. Eggs are laid from
the end of September until November. The eggs are olive-green when first
laid but soon turn black and shiny. Normally they are laid in the spaces
between the buds and the twigs but they may be found in any crevice
(Plate VII). When few, they are usually on the new shoots or adjacent two-
year-old shoots but, when numerous, they are also in fissures in the bark of
the older wood.

During the summer the black bean aphid colonizes many ornamental
plants and weeds such as docks, poppies, fat hen and thistles, but on some of

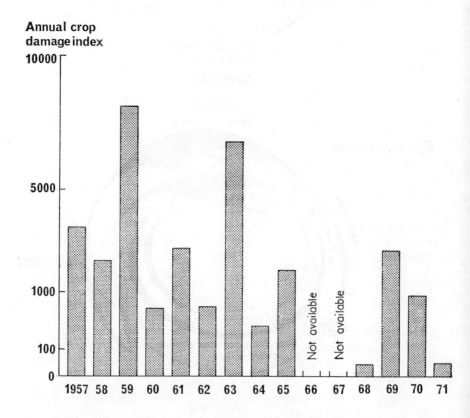

Fig. 7(*a*). Extent of crop damage by black bean aphids, 1957–71.
Note the tendency for crop damage to rise and fall in alternate years (see page 50).

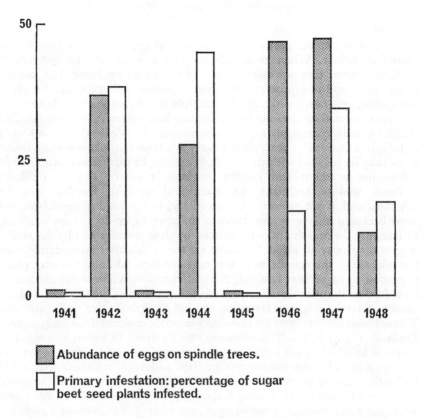

■ Abundance of eggs on spindle trees.

☐ Primary infestation: percentage of sugar
   beet seed plants infested.

Fig. 7(*b*). Abundance of black bean aphid eggs on spindle trees related to primary
infestation in seed plants, 1941–48.

Note the tendency for egg numbers and the percentage of infested plants to rise and
fall in alternate years (see page 50).

them other species of black aphid also occur, e.g., the permanent dock aphid, *Aphis rumicis* L., which does not migrate to beans, sugar beet or mangels. Several species of black aphid lay eggs on the spindle tree. The black aphid commonly found on the wild guelder rose (*Viburnum opulus* L.) is usually the viburnum aphid, *Aphis viburni* Scop. On the evergreen Japanese spindle, common in gardens in southern England, another black aphid, the Japanese spindle aphid, breeds throughout the winter in sheltered places. None of these aphids colonizes sugar beet or mangels in the field.

EPIDEMICS

Outbreaks depend partly on the number of eggs on spindle trees in the preceding winter. When these are few, there is usually no subsequent epidemic; when they are many, there is risk of an epidemic, but various factors may tend to suppress it, the most important being the weather. In a fine spring, eggs hatch early, aphids multiply rapidly on the spindle tree and migrate early and successfully to the summer host plants. A fine, dry summer enables aphids to multiply rapidly and migrate freely. By the second week in July, in a favourable year, almost every available host plant over extensive areas may be infested with thousands of aphids. In exceptional years, crops throughout southern and eastern England, northern France, Belgium, Holland, western Germany, Denmark and southern Sweden may be colonized and, if not protected by insecticide, loss in yield of sugar beet and sugar beet seed may be severe. More often, however, epidemics are localized. In late July the aphids are overwhelmed by their enemies and by the end of August are nearly all killed. So scarce are the aphids after an epidemic, that few migrate to the spindle tree; few eggs are then laid and the next year's migration to sugar beet is slight. Natural enemies decline in the absence of their food so that the remaining aphids prosper during the summer and the numerous return migrants lay many eggs on the spindle tree in the autumn. This explains the tendency for epidemics to occur in alternate years in eastern England (Figs. 7(*a*, *b*)). This tendency can be upset, however, by long, fine autumns in which small aphid populations tend to increase (e.g., 1947 and 1949), or by cold, wet summers (e.g., 1946 and 1948) which restrict aphid multiplication and predator increase during June and July so that the aphid population is maintained during August. For reasons of this kind, many eggs were laid every autumn between 1946 and 1950 giving a constant threat of epidemics, which fortunately did not always materialize.

Since 1948 British Sugar Corporation fieldmen have counted the number of black aphids per plant in sample fields at the end of each month from May to September inclusive; in addition, from 1957 to 1965 and since 1968 the damage caused to sugar beet has been estimated by each fieldman in his area each month from April to July. Years of greater and lesser damage alternated regularly in these periods until 1971 (Fig. 7(*a*)). In 1970 the infestation was large but it caused little damage because it was late; apparently the numerous predators produced in 1970 overwintered successfully and, aided by a cold wet June, prevented the build-up of an infestation in 1971. In July 1963, black aphids were more numerous on beet than in any of the previous 20 years but, because infestation was late, the damage caused was not as great as in 1959.

Forecasting black aphid epidemics is being attempted, but the practical value of such early forecasting is largely obviated by the spray warning scheme operated by the British Sugar Corporation, based on fieldmen's daily counts on the root crop during May and June (see page 61).

(see page 61)

CONTROL: ERADICATION OF SPINDLE TREES

Eradication of spindle trees has been suggested because the aphid is dependent on them for overwintering. Several objections to this proposal have been raised, one being that the spindle tree is not the only winter host and the aphid might successfully overwinter on alternative hosts or adapt to new ones. However, observations extending over many years indicate that the spindle tree is the principal winter host while the wild guelder rose and various exotic spindles, viburnums and other ornamental bushes planted in parks and gardens are of little importance. Moreover, eradication of spindle would not increase the likelihood of adaptation to new hosts, because every year most return migrants fail to find spindle trees and perish without colonizing new hosts. It has also been suggested that, like certain other aphids, the black bean aphid might overwinter in the asexual, summer phase; occasionally individuals do this, for instance in mangel clamps, but they are of no importance in initiating infestations in the root crop (compare with the peach-potato aphid, page 53). Another objection to eradication of spindle trees is the possibility that aphids might be carried in air currents from the Continent, but few would be likely to arrive before July, especially in the main beet-growing areas north of the Thames. Also, those interested in the conservation of our native plants object to spindle tree eradication.

The arguments for and against eradication are debatable but the practical problems would be great. To the casual observer the spindle tree appears uncommon or even rare, but it occurs, and in some places is abundant, on calcareous soils south of a line from the Wash to the Severn; it becomes scarcer further north and is almost absent north of a line from the Forth to the Clyde. Finding the bushes, removing them and repairing gaps left in hedgerows would be a Herculean task rendered more difficult by the vigour with which the bush sends up suckers from fragments of roots. The spindle is also a fairly common decorative shrub in gardens.

Eradication on a country-wide scale is impracticable, but removing bushes from the hedgerows of beet fields will decrease the risk of infestation, for the earliest infestations occur near concentrations of bushes. Cutting back during every other winter is also beneficial, because it destroys many overwintering eggs and leads to a type of growth that is not favoured by the return migrants in the autumn.

CONTROL: CROP CULTURE AND TIMING OF INSECTICIDE APPLICATION

Crop rotation offers no protection against aphids, because of the ease with which the winged migrants are transported in air currents. However, early-sown plants are more resistant to infestation and injury than late-sown plants; thick stands are also advantageous as the plants are generally less heavily infested than those in thin stands.

In epidemic years growers used to have difficulty in deciding whether and when to apply insecticides, but this difficulty is now largely obviated by the sugar factories' spray warning scheme. This warning is primarily for green aphids but applies also to black aphids (see page 61 for details). In general, the pest threatens damage if 10 per cent or more of the plants have colonies in the heart leaves in early June. The crop must then be closely watched, especially if the weather continues fine and warm. Where more than 25 per cent of plants are infested and the colonies are obviously increasing, the crop should be treated without delay. Another guide to the timing of treatment is that yield is reduced as soon as there are more than an average of about two aphids per leaf. To be fully effective, insecticide must be applied before the heart leaves become curled and the plants suffer a serious check. Treatment is usually undertaken at the end of June and in early July, but the timing differs from year to year and is usually later in the north. In late July, when the aphid colonies are overtaken rapidly by parasites and predators, treatment is no longer worth while.

Seed merchants advise growers when treatment of seed crops is necessary; here the progress of an infestation is more obvious because the black colonies spread and extend down the flowering stems. Mild attacks can be controlled by treatment of the crop margins, on which the infestation is usually more dense. If topping, where done, can be delayed until the first week in June and the tops collected and destroyed, many of the primary infestations are eliminated. In bad years, however, such measures are insufficient and the whole field must be treated with insecticide if yield and seed quality are to be maintained.

CONTROL: RECOMMENDED INSECTICIDES

Nicotine dust or fumigation was extensively used against this pest but has been replaced by systemic organophosphorus compounds. Materials such as DDT, gamma-BHC and trichlorphon should not be used, even when there are very few black aphids on a crop, because they are usually more destructive to aphid enemies than to the aphids themselves and can, therefore, actually cause an increase in infestation.

Systemic insecticides are recommended, either as sprays in not less than 20 gallons of water per acre, or as granules. Some systemic insecticides act at first by fumigant and contact action; all are absorbed by the plant and carried in the sap. Too much reliance should not be placed on absorption and translocation however: a good initial cover is essential. On small, rapidly growing plants, a spray persists for only a few days, but granules applied to large plants in late June persist for three or more weeks. Suitable spray materials are demephion at 3·6 oz of active ingredient (12 fl. oz of 30 per cent emulsifiable concentrate) per acre *or* demeton-S-methyl at 3·4 oz of active ingredient (6 fl. oz of 58 per cent emulsifiable concentrate) per acre *or* dimethoate at 4·8 oz of active ingredient (12 fl. oz of 40 per cent emulsifiable concentrate) per acre *or* ethoate-methyl at 4·8 oz of active ingredient (24 fl. oz of 20 per cent emulsifiable concentrate) per acre *or* formothion at 7·5 oz of active ingredient (16 fl. oz of 43 per cent emulsifiable concentrate) per acre *or* menazon at 6 oz of active ingredient (15 fl. oz of 40 per cent emulsifiable concentrate) per acre *or* oxydemeton-methyl at 3·4 oz of active ingredient (6 fl. oz of 57 per cent emulsifiable concentrate) per acre *or* phosphamidon at 3·2 oz of active ingredient (e.g., 16 fl. oz of 20 per cent emulsifiable concentrate) per acre *or* thiometon at 4 oz of active ingredient (16 fl. oz of 25 per cent emulsifiable concentrate) per acre. For larger plants, the recommended dose should be increased by at least 25 per cent, especially with dimethoate and menazon. The application of 16·8 oz of active ingredient per acre of disulfoton (14 lb of 7·5 per cent granules) or 16 oz of phorate (10 lb of 10 per cent granules), with the granules concentrated in bands over the rows, controls black aphids longer and is more effective than spraying, especially when the plants are heavily infested or late in the season; alternatively these amounts may be split between two applications and are then approximately equivalent to two sprays. Early treatment is essential to gain maximum benefit and treatment after the middle of July is unlikely to be worth while.

Infestations on seed crops can be controlled with any of the above sprays in at least 50 gallons of water per acre, applied just before the crop grows too tall for tractor-mounted sprayers. Band application to the foliage of at least 24 oz of active ingredient per acre of disulfoton or phorate granules will give good and lasting control of aphids if applied just before the plants exceed 2 ft in height. Infestations that develop after the crop is too tall to be treated by tractor-drawn machines can be treated from aircraft.

## Peach-Potato Aphid

The peach-potato aphid (*Myzus persicae* (Sulz.)) is the most serious pest of sugar beet, mangels, fodder beet and red beet. Unlike the black bean aphid, it is rarely numerous enough on beet to cause direct injury by its feeding; its importance is almost entirely due to its activity as a vector (carrier) of viruses, especially yellows viruses. These virus diseases depress yield and quality of beet and related crops to an extent depending on how early in its development the plant is infected. After several years of relative scarcity, yellows was particularly prevalent in 1957 and fairly prevalent in 1958–61 but it was much less prevalent in 1962–71.

### DESCRIPTION AND LIFE CYCLE

The wingless aphid is usually pale green but the young that are developing wings may be reddish-brown. The winged form has a shiny, black head and thorax while the abdomen is olive-green with a characteristic black patch on the upper surface (Fig. 10(*a*)). The life cycle (Fig. 6) differs from that of the black bean aphid in that the winter, especially a mild one, is passed as an immature or adult aphid, as well as in the egg stage. Aphids overwinter on field or garden brassica and chenopod crops, and on field, hedgerow and garden weeds, as well as in glasshouses or in mangel clamps. Where the winter is severe (e.g., in eastern Europe), eggs are the main overwintering stage. The woody winter host on which eggs are laid is the peach. Peach is not a host of beet viruses and the egg stage is free of virus; winged aphids leaving peach in the spring thus have to pick up virus from some infected plant before they can infect beet. Summer hosts include many weeds and crops.

The efficiency of the aphid in transmitting of virus deseases is related to its habits: the wide host range ensures survival throughout the year; the ability to overwinter as an aphid enables it to migrate, if the weather is favourable, at a much earlier date in spring than aphids that hatch from eggs on their winter host trees; the migrating aphid may already be infected with virus from the herbaceous host plant on which it has overwintered; the aphid is restless, moving readily from plant to plant and, although its powers of

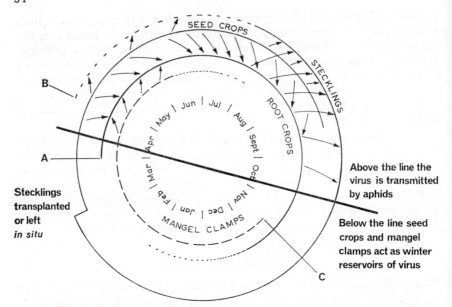

Fig. 8. Cycle of yellows and mosaic viruses transmission by aphids between seed crops, root crops and mangels

A. The sowing of sugar beet and mangel root crops begins in March. Virus infection can be brought to these crops by winged aphids from the seed crop and from mangel clamps. Virus transmission reaches its peak in July and dies away again. Steckling beds and mangel clamps can act as important winter reservoirs of virus until transmission recommences in spring.

B. Steckling beds of sugar beet, mangels and beetroot are sown in July and become infected with virus by aphid transmission (arrows) from the root and seed crops. The stecklings are later transplanted or grown on *in situ* as the seed crop of the following year.

C. Mangels are clamped in late autumn and early winter (dotted line). The clamps contain virus-infected roots and some persist until May or even June. Aphids, some of which may have overwintered in the clamps, carry the virus to the root crop.

reproduction are very great, large colonies are rarely formed. The peach-potato aphid is also intrinsically a better vector of viruses than is the black bean aphid, which usually plays only a small part in the spread of yellows viruses.

THE VIRUS CYCLE

The yellows and mosaic viruses are not seedborne; therefore, for their transmission, a source of the virus and the presence of aphids, especially winged aphids, is essential (Fig. 8). The aphids feed by sucking sap from infected plants and become contaminated with virus. Mosaic virus does not persist in an aphid for more than several hours, but peach-potato aphids contaminated with the beet yellows virus remain infective for several days and with beet mild yellowing virus for most or all of their lives. These facts explain the different distribution of the virus diseases on beet, mosaic usually being found only very near its source (particularly seed crops) while the yellows virus and especially the mild yellowing virus are more generally distributed.

Suitable conditions for aphid migration and multiplication occur from April to October in all the important beet-growing areas. On fine days winged aphids may be carried up to great heights by thermal up-currents and transported far on light winds before they settle upon the vegetation. Aphid migration ceases in winter but a reservoir of virus remains in seed crops of beet, mangels, fodder beet and red beet, in clamps of mangels, fodder beet and red beet, in overwintering groundkeepers, in garden crops of seakale beet and spinach beet, and in some weeds. Wild beet (*Beta vulgaris* L. ssp. *maritima* (L.) Thell.) may act as an overwintering source of virus, but it is relatively unimportant as the peach-potato aphid rarely overwinters on it. Because severe winters kill many virus-infected plants and aphids, the incidence of yellows virus is related to the number of days in January, February and March when the temperature falls below 0°C. The average temperature for April should also be taken into account in forecasting the likely incidence of virus infection; May to July weather seems to have surprisingly little effect but may, of course, markedly influence the development of black aphid infestation. Thus virus incidence can be forecast approximately and recommendations for insecticide treatment can be made with more assurance (see page 61).

VIRUS SYMPTOMS

In June, symptoms of virus yellows in sugar beet (see also Bulletin 142 and Advisory Leaflet 323) appear ten to fourteen days after injection of the virus during the feeding of an infective aphid. Later in the year, and in dull weather, the time between infection and appearance of symptoms may be as great as four to six weeks in older plants. Yellowing is first observed on the tips of middle-aged leaves but this gradually extends until all the outer leaves are affected and only the heart leaves remain green. Yellowed leaves are thickened and brittle, crackling when crushed in the hand. Two viruses cause yellowing, each producing distinguishable symptoms: (1) beet yellows virus (BYV)—the centre leaves are often 'vein-etched' when the infection is recent, there are invariably small, brown or red necrotic spots on the old, pale yellow leaves when the infection is of long standing, and the yellowed leaves are not commonly attacked by secondary fungal pathogens; (2) beet mild yellowing virus (BMYV)—infected plants have orange-yellow leaves with no brown or red necrotic spots, and the leaves are more often attacked by secondary fungal pathogens such as *Alternaria* spp. than leaves on plants infected with BYV. The symptoms may be confused with yellowing caused by capsids (page 63), red spider mites (page 65), downy mildew (page 104) or deficiency diseases (see Bulletin 142).

The leaves of plants infected with beet mosaic virus are finely mottled with light and dark green. When leaves are held up to the light, the mottle is seen as small, light green rings on a darker background, or dark green bands along the veins with lighter areas between. Infected plants are not stunted or deformed, and their leaves do not become necrotic. The symptoms are most pronounced in the heart leaves and usually fade as the leaves grow older (see Bulletin No. 142).

CONTROL BY CROP MANAGEMENT

Experiments have shown that considerable advantage is gained by early sowing and dense plant stands. At Broom's Barn Experimental Station on

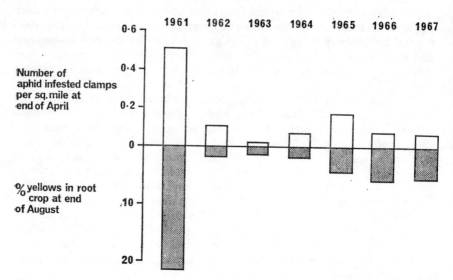

Fig. 9. Relationship between the number of aphid-infested clamps per square mile at the end of April and the incidence of yellows viruses in the root crop at the end of August (specific field counts)

21st July 1964, there were an average of 1·3 wingless green aphids per plant on plots sown on 2nd April, but 2·9 per plant on plots sown on 28th April, although all plots had 25,000 plants per acre. This difference was accentuated on plots with greater or smaller plant populations, viz., 0·8 wingless green aphids per plant sown on 2nd April and singled to 42,000 plants per acre, but 4·5 per plant sown on 28th April and singled to 16,000 per acre. By 15th September 2 per cent and 18 per cent, respectively, of these plants were infected with virus. Early sowing and thick stands discourage aphid infestation and decrease virus infections; in addition, the older the plants are when they become infected, the less the effect on their yield and quality. Plant breeders are selecting lines of sugar beet that are better able to tolerate infection with yellows viruses and one monogerm variety is available commercially; such a variety is recommended by the National Institute of Agricultural Botany for areas where yellows is prevalent. Considerable improvement in the yield, quality and appearance of late-sown crops may result from spraying with a systemic insecticide, but the same improvement in yield and quality can usually be obtained by early sowing, although the early-sown crop may appear more yellowed at the end of the season.

CONTROL BY HYGIENE

Insecticides decrease aphid numbers effectively but usually cannot prevent virus-carrying aphids infecting the plants with virus before they absorb enough insecticide to be killed. Efforts to decrease or eliminate the winter reservoirs of virus (see page 55 and Figs. 8 and 9) are well worth while. Beet crowns and unharvested roots often survive and send up seed shoots in the following crop; in cereals they are usually unimportant, being screened or killed by herbicide, but when growing in crops such as potatoes or carrots they can be an important source of virus. Sugar beet seed crops and mangel clamps are major over wintering sources of virus; much effort has been devoted to decreasing their importance as virus sources in the spring and this has been successful with seed crops.

Eliminating infection from seed crops begins with the stecklings, which are either (1) sown in April under a cover crop (usually barley), or (2) sown in July-August in open beds for transplanting later, or (3) sown in July-August with or without cover, and grown on *in situ*, usually in areas isolated from the districts where the root crop is grown. Insecticide treatments to protect the stecklings from aphids and infection with yellows viruses are obligatory and are specified in the contract between the British Sugar Corporation and the seed merchants (see page 60 for insecticide recommendations). When grown under a cover crop, especially if sown in April, the stecklings are unlikely to need protection with insecticide until the cover crop is harvested; stecklings grown without cover must be protected with systemic insecticide from seedling emergence until late October.

Under the terms of the seed merchants' contract with the British Sugar Corporation, representatives of both inspect in October all stecklings growing in root crop areas. Steckling beds with less than one per cent infection are certified as satisfactory, those with more than ten per cent infection are condemned, and those with infection varying from one to ten per cent are used only if absolutely necessary to complete the acreage of seed needed. This scheme of inspection also voluntarily covers a sample of mangel, fodder beet and red beet beds. The increased yield of seed from healthy stecklings is striking; it compensates the seed grower for the inconvenience that the scheme causes and greatly safeguards the health of neighbouring root crops. To prevent the spread of viruses within seed crops and incursions from the root crop, aphid control may be needed in May or June (see page 60 for insecticide recommendations).

Mangel clamps create a problem. On average, about one-third of the mangel clamps remaining in late April are infested with aphids; nearly all of these contain the mangold aphid (*Rhopalosiphoninus staphyleae* (Koch)) and about half contain the peach-potato aphid. Most of the mangels are infected with yellows viruses. Winged aphids leave the clamps from mid-April onwards and their effects are often obvious later in the season as a greater incidence of yellows around clamp sites than elsewhere, although some winged aphids from clamps fly a long way and infect occasional plants over a wide area. The number of aphid-infested clamps per square mile in late April 1961–67 paralleled the incidence of yellows in beet fields at the end of August in those years (Fig. 9); the incidence of infested clamps per square mile has not been recorded since 1967. Whenever possible, clamps should be cleared and the mangels fed to stock before the beginning of April, as the aphids increase rapidly in the clamps during April and May. Roots no longer required should be destroyed, the straw and other debris burnt and the clamp site ploughed to destroy any shoots or weeds growing. Where clamps must be kept until later and are infested with aphids, control is difficult. Fumigation with methyl bromide controls aphids excellently but is too expensive. *If* the mangels are surplus and are *not* to be fed to stock, liberal treatment on the surface of the mangels with gamma-BHC or with one of the

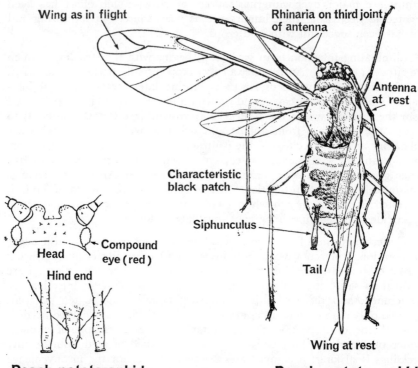

Wing as in flight

Rhinaria on third joint of antenna

Antenna at rest

Characteristic black patch

Siphunculus

Head

Compound eye (red)

Hind end

Tail

Wing at rest

**Peach-potato aphid: wingless female**

**Peach–potato aphid: winged adult**

## Wingless green aphids common on beet
### (relative size of adults)

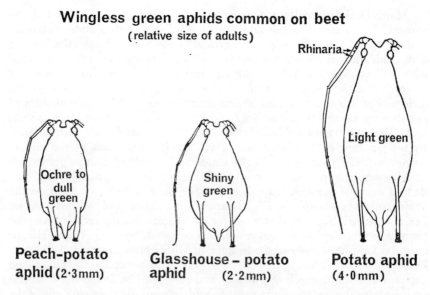

Rhinaria

Light green

Ochre to dull green

Shiny green

**Peach-potato aphid** (2·3 mm)

**Glasshouse – potato aphid** (2·2 mm)

**Potato aphid** (4·0 mm)

Fig. 10(a). The peach-potato aphid and other green aphids common on beet

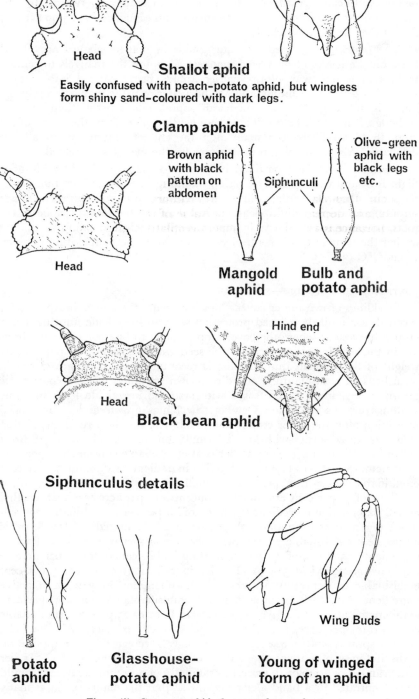

**Shallot aphid**

Easily confused with peach-potato aphid, but wingless form shiny sand-coloured with dark legs.

Hind end

Head

**Clamp aphids**

Brown aphid with black pattern on abdomen

Olive-green aphid with black legs etc.

Siphunculi

Head

**Mangold aphid**   **Bulb and potato aphid**

Hind end

Head

**Black bean aphid**

**Siphunculus details**

Wing Buds

**Potato aphid**   **Glasshouse-potato aphid**   **Young of winged form of an aphid**

Fig. 10(b). Common aphids that reproduce on beet

systemic insecticides recommended for aphid control (see below), preferably granular disulfoton or phorate, usually kills the aphids. The covering should be removed before treatment and replaced immediately afterwards, and the mangels must be left to rot and then spread and ploughed in; on no account should they be fed to stock.

The overwintering peach-potato aphids in mangel clamps are placed there on the leaves of harvested mangels. Topping the mangels to remove the leaves is one way of preventing aphids being carried into clamps. Removing leaves by running a forage harvester over the mangel crop before lifting is not as good as careful hand-topping but is better than nothing. Although it may not be possible to remove the leaves from the whole of the crop, that part likely to remain after the end of March can usually be defoliated. A separate clamp is necessary for the mangels so treated.

Mangels should not be lifted before mid-October and preferably after slight frost and when wet. The clamp should not be covered too thoroughly too soon, otherwise heating will occur. High clamp temperatures encourage sprouting and aphid multiplication, and decrease the feeding value of the roots. In Denmark considerable importance is attached to clamp ventilation during the winter; this is controlled so as to keep the temperature between 1° and 4°C (34°–39°F).

CONTROL BY INSECTICIDES—SEED CROP

Stecklings grown under cover crops, especially those sown in April under a cereal, are unlikely to need protection with an insecticide until the cover crop is harvested; where the cover crop is likely to be poor when the beets are in the seedling stage, treating the seed with menazon at 5 per cent by weight of 80 per cent dust dressing is recommended. Stecklings grown in open beds for transplanting, and direct-sown stecklings without cover to be grown *in situ*, must be protected with systemic insecticide from the time seedlings emerge until late October. Stecklings grown under cover need protection after the cover crop is harvested. The standard recommendation is at least three sprays with liquid formulations of demephion at 3·6 oz of active ingredient (12 fl. oz of 30 per cent emulsifiable concentrate) per acre *or* demeton-S-methyl at 3·4 oz of active ingredient (6 fl. oz of 58 per cent emulsifiable concentrate) per acre *or* dimethoate at 4·8 oz of active ingredient (12 fl. oz of 40 per cent emulsifiable concentrate) per acre *or* ethoate-methyl at 4·8 oz of active ingredient (24 fl. oz of 20 per cent emulsifiable concentrate) per acre *or* formothion at 7·5 oz of active ingredient (16 fl. oz of 43 per cent emulsifiable concentrate) per acre *or* menazon at 6 oz of active ingredient (15 fl. oz of 40 per cent emulsifiable concentrate) per acre *or* oxydemeton-methyl at 3·4 oz of active ingredient (6 fl. oz of 57 per cent emulsifiable concentrate) per acre *or* phosphamidon at 3·2 oz of active ingredient (e.g., 16 fl. oz of 20 per cent emulsifiable concentrate) per acre *or* thiometon at 4 oz of active ingredient (16 fl. oz of 25 per cent emulsifiable concentrate) per acre, in at least 20 gal of water per acre. Alternatively, any two spray applications except the first may be replaced by one application of disulfoton or phorate at about 16 oz of active ingredient (14 lb of 7·5 per cent granules or 10 lb of 10 per cent granules) per acre; the treatment should be concentrated in a band along the row so that the plants retain as many granules as possible. Adequate and timely protection of the very young seedlings is difficult by foliage treatment because sprays do not persist and granules are not retained, but they can be protected with menazon seed dressing (as above), which kills green aphids until at least the first pair of true leaves has expanded. Subsequent protection can be obtained from sprays or granules as above.

Aphid infestations of the seed crop need controlling to prevent winged aphids leaving the crop and spreading viruses to root crops and, if the infestation is heavy, to prevent damage to the seed crop. Control is achieved by spraying with demephion *or* demeton-S-methyl *or* dimethoate *or* ethoate-methyl *or* formothion *or* menazon *or* oxydemeton-methyl *or* phosphamidon *or* thiometon, at the rates of active ingredient recommended for the stecklings but in at least 50 gal of water per acre, just before the crop grows too tall for tractor-mounted sprayers. Band application to the foliage of disulfoton or phorate granules at a minimum of 24 oz of active ingredient per acre gives good and lasting control of aphids when applied just before the plants exceed two feet in height. Crops that become infested after they are too tall to be treated by tractor-drawn machines can be treated from aircraft.

CONTROL BY INSECTICIDES—ROOT CROP

Removing the sources from which yellows viruses spread in spring diminishes the number of plants in root crops that become infected with virus early on, but inevitably some infective aphids introduce and spread the disease. Infections can be delayed by well-timed treatment with systemic insecticide. The time at which aphids invade root crops differs greatly in different years and different districts, so neither seed treatment nor soil treatment at sowing is recommended as a routine. However, for later sowings in the south-east of England and areas where aphids and yellows usually arrive early, seed treatment with menazon (80 per cent dust applied at the rate of 5 per cent by weight of raw seed) is justified. Unfortunately menazon cannot be incorporated in pelleted seed.

Timing of foliage treatments is important and the British Sugar Corporation warns growers by postcard when the time seems opportune for treatment. The issue of spray warnings is decided on the basis of local and general information about the development of aphid populations. To obtain local information, each day each fieldman counts the numbers of wingless green aphids per plant in four fields and posts the results to his factory. To obtain general information, each week each factory submits the results of all counts to Broom's Barn Experimental Station, where it is summarized and circulated to all beet sugar factories, the Agricultural Development and Advisory Service and others who are interested. In the areas where yellows viruses are prevalent, an average of about 0·25 *wingless* green aphids per plant justifies the warning cards, but the prevalence of sources of yellows viruses, aphid activity, stage of growth of the crop, and the weather are also taken into consideration. In early seasons aphids may infest the crop in the seedling stage, but then treatment is recommended only where the infestation is heavy. Spraying with any of the materials listed on page 52 rapidly checks the aphid infestation, and the insecticidal effect persists for a week or more. Band spraying is effective when done carefully and is economical. A second spray two or three weeks after the first may be worth while but a third is rarely needed. Disulfoton or phorate granules applied to the foliage are toxic to aphids for longer, control yellows more effectively and increase yield more than do sprays. The granule applicator can be mounted on the steerage hoe, thus combining insecticide treatment with inter-row cultivation; the granules should be directed in a band the width of the plants so that as many as possible are retained by the leaves. Use disulfoton or phorate at about 16 oz of active ingredient (14 lb of 7·5 per cent granules or 10 lb of 10 per cent granules) per acre as one, or a split, application. Granules should not be used on plants before singling, especially during dry weather.

## OTHER APHIDS

Winged aphids of many species are abundant in June and July and some alight, largely by chance, on the beet crop. Green aphids other than the peach-potato aphid may thus be found on the crop, especially in the spring and early summer. The potato aphid (*Macrosiphum euphorbiae* (Thos.)) is often found on beet, producing young wingless aphids which are usually able to feed and grow to maturity. It is a much larger aphid than *Myzus persicae*, and the wingless form has an elongated body and comparatively long antennae and siphunculi. Pink individuals are not rare. Another species often found on beet at this time of year is the glasshouse-potato aphid (*Aulacorthum solani* (Kltb.)), the wingless form of which is shiny green (appearing wet) with long antennae having dark joints. Winged individuals of the shallot aphid (*Myzus ascalonicus* Donc.) may also be found; they are extremely difficult to distinguish from *M. persicae* although they tend to fly earlier in the season and deposit only very few young on beet plants. The young of *M. ascalonicus* are sand-coloured with dark legs and antennae, never green. Winged aphids of many other species alight on beet, but they do not stay for long and seldom produce young there. These aphids do not play an important part in the transmission of viruses as they are less efficient vectors and they overwinter less commonly on virus host-plants; they are usually not numerous enough to cause direct damage by feeding. The presence of winged green aphids on the beet plants in May and June indicates that migration from other plants is occurring but does not necessarily indicate the need for insecticide treatment; the invasion may be of aphids from sycamore for instance. The British Sugar Corporation's spray warning is based on numbers of *wingless* green aphids per plant, which indicate that colonization by *M. persicae* and probably some of the above aphids is occurring and that an insecticide should be applied before numbers increase too much and the aphids spread viruses within the crop. Fig. 10 illustrates the diagnostic features of the three green aphids that are common on beet plants in the field, and also those found in mangel clamps. The mangold aphid (*Rhopalosiphoninus staphyleae* (Koch)) is the commonest aphid in mangel clamps and is sometimes very numerous. It is a dull olive-green or brown aphid and is able to transmit some beet viruses, but it is not found on the foliage of beet plants in the field as it is a root-feeding species. For its control in mangel clamps see page 57.

## FURTHER READING ON APHIDS

DUNNING, R. A. and WINDER, G. H. (1965). The effect of insecticide applications to the sugar beet crop early in the season on aphid and yellows incidence. *Pl. Path.* **14,** Suppl. 30–6.

EMDEN, H. F. VAN, EASTOP, V. F., HUGHES, R. D. and WAY, M. J. (1969). The ecology of *Myzus persicae*. *A. Rev. Ent.* **14,** 197–270.

HEATHCOTE, G. D. and COCKBAIN, A. J. (1964). Transmission of beet yellows virus by alate and apterous aphids. *Ann. appl. Biol.* **53,** 259–66.

HEATHCOTE, G. D. and COCKBAIN, A. J. (1966). Aphids from mangold clamps and their importance as vectors of beet viruses. *Ann. appl. Biol.* **57,** 321–36.

HEATHCOTE, G. D., DUNNING, R. A. and WOLFE, MARIE D. (1965). Aphids on sugar beet and some weeds in England, and notes on weeds as a source of beet viruses. *Pl. Path.* **14,** 1–10.

HULL, R. (1961). The health of the sugar beet crop in Great Britain. *Jl R. agric. Soc.* **122,** 101–12.

HULL, R. (1968). The spray warning scheme for control of sugar beet yellows in England. Summary of results between 1959–66. *Pl. Path.* **17,** 1–10.

JONES, Margaret G. (1969). The bean and beet aphid *Aphis fabae* Scop. *Sch. Sci. Rev.* No. 172, 549–57.

KENNEDY, J. S. and STROYAN, H. L. G. (1959). Biology of aphids. *A. Rev. Ent.* **4,** 139–60.

STROYAN, H. L. G. (1952). The identification of aphids of economic importance. *Pl. Path.* **1,** 9–14, 42–8, 92–9 and 123–9.

## CAPSIDS

Sugar beet can be damaged by different species of capsid (Miridae) when it is in different stages of growth. The tarnished plant bug (*Lygu rugulipennis* Popp.) is the chief species that attacks beet in the seedling stage. It over-winters as a winged adult in hedgerows and similar cover and migrates into beet fields early in the year. The bugs are greenish-brown and about $\frac{1}{4}$ in. long; they are larger, more active and more robust than aphids, which they otherwise resemble. Like aphids they pierce plant tissue with their elongated mouthparts, inject saliva and suck the sap. When disturbed they run or fly from plant to plant. They feed on any over-ground part of the seedling but favour the growing point. This has disastrous results, for their saliva is toxic and kills the growing point; subsequently the heart tissue often swells and the cotyledons thicken and enlarge (Fig. 4(a)). Eventually fresh growing points arise and the plant develops a multiple crown, but not before suffering a setback from which it never fully recovers. Symptoms appear some time after the damage has been caused and too late for insecticide treatments. Affected plants are easily distinguished and should be removed during hand singling if this is practised. Eggs are inserted into the plants but, as only a small part of the egg protrudes, they are not easily found. Wingless young emerge from the eggs after two or three weeks and feed on the maturing plants but cause little further damage; the next generation of adults may feed on and injure stecklings sown in July or August.

The species most commonly found in beet crops in summer are the common green capsid (*Lygocoris pabulinus* (L.)) and the potato capsid (*Calocoris norvegicus* (Gmel.)). The eggs of both overwinter on woody hedgerow plants. The young are green and wingless but larger and much more active than aphids. They walk from the hedgerows into adjoining beet crops and feed upon stems and leaves, injecting toxic saliva which kills cells and causes dead spots around the stab wounds (Fig. 4(c)). Possibly because of this toxicity, they do not transmit virus diseases. Tearing and distortion may occur as the leaves grow, producing irregular holes of variable size which give the leaf a tattered, mis-shapen appearance (Fig. 4(c)). This form of injury occurs near hedgerows and orchards and is most noticeable in small fields surrounded by overgrown hedges, which provide the overwintering sites for the eggs. Another form of injury arises when capsids feed on the leaf veins of older plants; one or more swollen stab wounds can usually be found on the central vein or on one of the main lateral veins, while beyond the wounds the leaf lamina puckers and the leaf tip becomes distorted and yellowed (Plate VI and Fig. 4(c)). The yellowing closely resembles that resulting from yellows virus infections, but the true cause is indicated by the presence of capsids in the crop, their feeding punctures and the fact that damage is adjacent to hedgerows.

Spraying with DDT at 16 oz of active ingredient (e.g., 4 pints of 20 per cent emulsifiable concentrate) in at least 20 gal of water per acre controls capsids; usually only the margins of the field need treatment. Phorate granules, at the rate recommended for aphids (see page 62), are also effective.

FURTHER READING

Dunning, R. A. (1957). Mirid damage to seedling beet. *Pl. Path.* **6,** 19–20.

Southwood, T. R. E. (1955). The nomenclature and life-cycle of the European tarnished plant bug, *Lygus rugulipennis* Poppius (Hem., Miridae). *Bull. ent. Res.* **46,** 845–48.

## Thrips*

Injury by thrips, usually the cabbage thrips (*Thrips angusticeps* Uzel), is uncommon but occurred widely in 1963 and 1970. The thrips overwinter in the soil as brachypterous (almost wingless) adults and are found on the seedlings in April and May feeding mainly on the curled heart leaves. When the leaves unfold and expand, they are elongated and even strap-like, roughened, with irregular and partially reddened or blackened margins and tips (Fig. 4(a)). Small, silvery lesions on the leaf surfaces are also evident. Control may be obtained by spraying with DDT at 16 oz of active ingredient (e.g., 4 pints of 20 per cent emulsifiable concentrate) per acre in at least 30 gal of water.

FURTHER READING

Bonnemaison, L. and Bournier, A. (1964). Les thrips du lin: *Thrips angusticeps* Uzel et *Thrips linarius* Uzel (Thysanoptères). *Annls Épiphyt.* **15,** 97–169.

Gough, H. C. (1955). *Thrips angusticeps* Uzel attacking peas. *Pl. Path.* **4,** 53.

Morison, G. D. (1947–49). Thysanoptera of the London area (London thrips). *Lond. Nat.* 1946, No. 26, Suppl. 1–36; 1947, No. 27, Suppl. 37–75; 1948, No. 28, Suppl. 77–131. (Identification and short notes on all British species of thrips. Re-issued as 'London Naturalist' Reprint No. 59.)

---

* See also Advisory Leaflet 170, available from the Ministry (p. 108).

## Leafhoppers

Next to aphids, lea.hoppers (Plate II) are the most important agents transmitting viruses of crop plants; they transmit the curly top virus of sugar beet
in North America. However, leafhoppers have not yet been found to transmit
viruses of sugar beet in Great Britain, although one or two species are often
very abundant. When disturbed, they fly from leaf to leaf. They cause little
harm to beet in Britain and control measures are unnecessary. The common
froghopper (*Philaenus spumarius* (L.)) (Plate II) is similar and also occurs in
sugar beet crops. The young stage of this froghopper is the well-known
cuckoo-spit insect which sometimes twists beet leaves in the root and seed
crops.

## Red Spider Mite*

In the autumn after exceptionally long spells of fine, dry weather, the
underside of sugar beet leaves may be colonized by a red spider mite,
most usually *Tetranychus urticae* (Koch). Affected leaves become yellow
between the veins. Yellowing usually appears first on plants around the
margins of the field, presumably by colonization from host plants existing
there. The feeding of the mites causes shallow depressions in which eggs,
young and adult mites can be seen with a hand lens. The mites also produce
silken threads which form a loose web; soil or dust thrown against the underside of the leaves is caught in the web and tends to adhere, whereas it readily
falls away when mites are absent. Attacks may be confused with other
causes of yellowing (see page 18).

---

* See also Advisory Leaflet 226, available from the Ministry (p. 108).

# Eelworm Pests

UNSEGMENTED round worms, or nematodes, occur in almost all natural environments. Some are parasites of animals, including man; others, commonly known as eelworms, are plant parasites or are free-living in the soil. Most eelworms are about $\frac{1}{25}$ in. long and only just visible to the naked eye, so that a lens or a microscope is necessary to find and study them. Like insects and their allies, their activity is greatly affected by temperature, so they become less active in winter. They also require a moist environment, so that in dry soil or during July and August there may be insufficient moisture for them to move, feed and reproduce.

All plant parasitic eelworms have piercing and sucking mouthparts (stylets), similar in action to the mouthparts of sucking insects but different in structure and far smaller. Like sucking insects, the saliva they inject while feeding may kill cells or induce the growth of galls. Because they are so small, great numbers are usually needed to cause serious harm to a crop, but three kinds, i.e., dagger eelworms (*Xiphinema* spp.), needle eelworms (*Longidorus* spp.) and stubby-root eelworms (*Trichodorus* spp.), transmit soil-borne viruses whose effects on plants may be much more serious than the direct effect of feeding.

Cyst eelworms and root-knot eelworms are specialized, internal parasites of plant roots and have life cycles as in Fig. 12. Cyst eelworms are very persistent and are readily transported in the cyst stage. The free-living or migratory soil eelworms (which include the virus carriers) and the stem eelworm have simple life cycles as in Fig. 11. Stem eelworms invade plants at or just below soil level and multiply in stem and leaf tissues; they may be carried well above soil level by plant growth. The last larval stage is able to survive drying and is readily transported. Thus, although eelworms can move only very short distances in soil or in plants, they are easily carried over great distances in soil or in farm produce. The eelworms that harm beet and the injuries they cause are as follows:

*Internal, migratory species*
Stem eelworm .. .. larvae burrow into stem and leaf petioles in spring, causing galling, malformation and occasionally blind seedlings. In autumn they invade the crown and initiate rotting.

*Internal, sedentary species*
Beet cyst eelworm or beet eelworm — larvae burrow into rootlets and injure or kill them during growth and maturation to the cyst stage.

Root-knot eelworms .. larvae burrow into rootlets and injure them during maturation inside the galls they produce.

*External, migratory species*
Free-living eelworms .. feed externally on rootlets, stunting growth and sometimes infecting the plant with viruses.

Stem eelworms may attack a crop in small numbers initially and then, like aphids, multiply sufficiently in one season to cause serious injury. Other eelworms that attack beet multiply in preceding host crops, persist over winter and begin to attack seedling root systems soon after germination. When numbers are great enough, the root system fails to grow properly and is unable to take up sufficient water and nutrients to support plant growth. Because eelworms cannot migrate more than limited distances, numbers and injury are usually related to past cropping.

FURTHER READING

JONES, F. G. W. and JONES, M. G. (1964). *Pests of field crops.* London, Edward Arnold.

SOUTHEY, J. F. (Ed.) (1965). Plant nematology. *Tech. Bull. Minist. Agric. Fish. Fd, Lond.* No. 7. H.M.S.O.

## BEET STEM EELWORM*

Crown canker caused by stem eelworm (*Ditylenchus dipsaci* (Kühn) Filipjev), known in Europe for about sixty years, was first recognized in England on mangels in 1927 and on sugar beet in 1936. It is now known that the

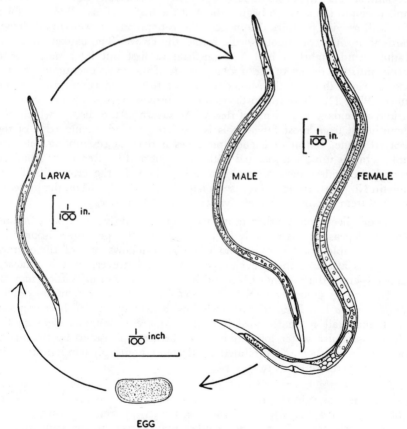

Fig. 11. Life cycle of the stem eelworm, *Ditylenchus dipsaci*

* See also Advisory Leaflet 178, available from the Ministry (p. 108).

eelworm also bloats and distorts leaf stalks, and sometimes kills the growing point of seedlings.

Stem eelworms are invisible to the naked eye within the tissues of the plants attacked but are just visible when suspended in water. The life cycle is simple (Fig. 11). The worms feed and multiply within the tissues of their host plants; eggs, young and adult worms occur together. Relatively few eelworms are present in galls on leaf stalks, in bloated leaves, or in invaded growing points of seedlings but, in the autumn, eelworms are usually abundant in cankered crowns, especially near the advancing edge of the rot.

When beet is attacked in the early cotyledon stage, the growing point is invaded and destroyed, and the cotyledons become malformed, twisted and thickened. Seedlings may be killed, but usually they survive and axillary growing points develop to produce a multiple-crowned beet with many small, distorted leaves (Fig. 4(d)). These symptoms are very similar to those produced by capsids (page 63) and sometimes by other pests, so the eelworm should be identified by a specialist. Eventually the plants grow more normally but remain stunted compared with healthy plants. Attacks occurring after the early seedling stage cause galls or bloating of the leaves and petioles (Fig. 4(c)); the plant suffers little or no damage and recovery is usually rapid. The crown canker seen in the autumn begins amongst the leaf scars, usually in the form of raised, greyish pustules. From here the rot spreads outwards and downwards to form a continuous girdle round the crown. The canker, which is granular and slightly raised, is superficial at first but later may extend deeply into the tissues of the upper portion of the root; eventually the rot extends right through the shoulder so that the crown comes away when pulled (Plate IX). The rotten tissue is dark brown and spongy and contains various secondary organisms of decay. A feature of the disease which distinguishes it from most fungal rots is the irregular advancing edge of the decayed tissue. Boron deficiency produces a dry rot resembling eelworm canker, but is usually associated with heart rot and is often more localized around the widest part of the root rather than on the crown itself (see Bulletin No. 142). In doubtful cases, diagnosis must be based on the presence of living stem eelworms at the advancing edge of the decay.

Besides beet, many other crops may serve as hosts, e.g., oats, onions, beans and peas. It has recently been shown that this pest is seed-borne by beans. Trouble sometimes occurs when beet follows one of these crops which was attacked the previous year. Often, however, beet is attacked following crops that are not hosts and here the source of infestation may be weeds, such as chickweed (*Stellaria media* (L.) Vill.) and common orache (*Atriplex patula* L.). Even when beet follows a heavily infested host crop, an attack may fail to materialize if conditions are unsuitable. Because the eelworms require water for activity, the disease is favoured by periods of cool, moist weather and inhibited by drought and high temperatures; its incidence therefore tends to be seasonal.

An early attack, causing multiple-crowned plants, is usually less severe than crown canker in the autumn. Multiple-crown plants are not seen later if blind seedlings (i.e., with dead growing points) are removed during hand singling. Yield losses from crown canker may exceed 10 per cent and, in addition, difficulties may be experienced in topping. No chemical control measures are available. Beet should not be grown after oats, beans or onions heavily attacked by stem eelworm. Crops with many cankered crowns in the autumn should be lifted early and delivered to the factory as soon as possible. Cankered roots are unfit for clamping.

FURTHER READING

DUNNING, R. A. (1957). Stem eelworm invasion of seedling sugar beet and development of crown canker. *Nematologica* **2,** Suppl. 362–8.

GRAF, A. and MEYER, H. (1971). Importance des pourritures des collets de betteraves provoquées par le *Ditylenchus dipsaci* en Suisse et solution de ce problème. *Int. Inst. Rech. Betteravier, 34th Winter Congress, Brussels.*

HOOPER, D. J. (1971). Stem eelworm (*Ditylenchus dipsaci*), a seed and soil-borne pathogen of field beans (*Vicia faba*). *Pl. Path.* **20,** 25–7.

STURHAN, D. (1969). Das Rassenproblem bei *Ditylenchus dipsaci. Mitt. biol. BundAnst. Ld-u. Forstw.* No. 136, 87–98.

## BEET CYST EELWORM*

Frequent cropping of land with sugar beet or mangels may lead to a condition known as 'beet-' or 'mangel-sickness' in which the plants are stunted and the yield greatly decreased (Plates VIII and XI). This condition is caused by the beet cyst eelworm or beet eelworm (*Heterodera schachtii* Schm.), a common pest in Europe and the United States and one of the major problems wherever beet is cultivated intensively. In Britain the pest is firmly established on the peat soils of the Fens and the Isle of Axholme and is slowly becoming more widespread elsewhere.

Enforcement of a suitable crop rotation (i.e., one with an adequate interval between host crops) in the known infested areas (see page 78 *et seq.*) and encouragement of similar rotations elsewhere have lessened the rate of spread and avoided serious yield losses. Rotation, however, cannot prevent spread, and the rate of spread must increase as more land becomes infested and provides more opportunities for the dissemination of eelworm cysts. The areas likely to be affected first are those where beet is grown frequently, or where beet and other host crops are grown in the same rotation. The pest often appears first on smallholdings and small farms where beet, mangels and brassicas are cropped in strips. It is often also associated with peaty or light land and high water tables, as in the Fens and some river valleys, but is not yet prevalent in the silt fens. Fig. 14 shows the known distribution of beet cyst eelworm in England and Wales up to 1970; it has also been reported from Scotland and Ireland. Fig. 15 indicates the infested areas in the Fens. Some of the scattered infestations in Fig. 14 are sewage farms which used to dispose of sewage effluents by land irrigation. The land was cropped frequently with mangels and brassicas, and sometimes with sugar beet, in the intervals between flooding. Fortunately, few such farms remain. Other isolated infestations are on what might be described as 'root' fields, which for convenience have been intensively cropped with susceptible roots. Amongst these fields is 'Barnfield', site of one of the classical experiments at Rothamsted Experimental Station, where mangels or sugar beet had been

---

* See also **Advisory Leaflet** 233, available from the **Ministry** (p. 108).

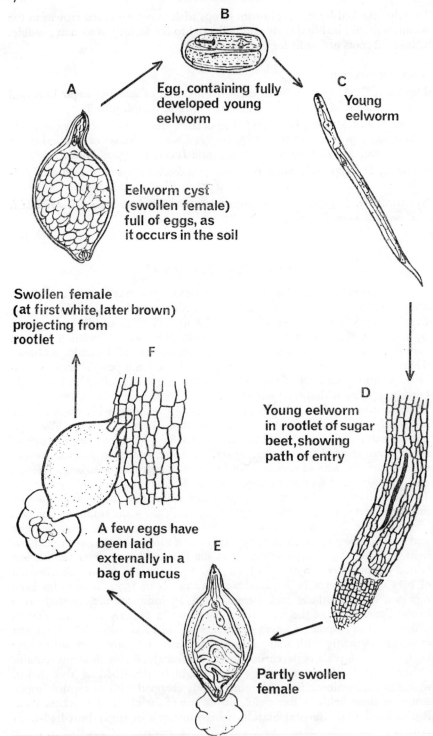

B

A

Egg, containing fully
developed young
eelworm

C

Young
eelworm

Eelworm cyst
(swollen female)
full of eggs, as
it occurs in the soil

Swollen female
(at first white, later brown)
projecting from
rootlet

F

D

Young eelworm
in rootlet of sugar
beet, showing
path of entry

A few eggs have
been laid
externally in a
bag of mucus

E

Partly swollen
female

Fig. 12. Life cycle of the beet cyst eelworm, *Heterodera schachtii*

grown continuously for over 80 years. Apart from the peat fens, other important infested areas are in Norfolk, Suffolk, the Holland and Lindsey Divisions of Lincolnshire, and the Isle of Axholme, Lincolnshire.

LIFE CYCLE

The eelworm lives entirely below ground. It has a resting stage or cyst from which young eelworms begin to emerge in the spring (Fig. 12). Emergence is greatly increased by the presence of substances given out by the roots (root diffusates) of sugar beet and other host plants. The larval eelworms, invisible to the naked eye, invade the rootlets of the host plant. As the eelworms grow, they swell and lose their worm-like appearance. Eventually, the cortex of the rootlet can no longer contain the eelworms and the skin of the root ruptures so that they become external, remaining attached by the head. At this stage both males and females become visible as small, white objects attached to the roots (Plate VIII). The males soon break free and regain their worm-like shape, but the females remain attached and continue to grow. A small bag of mucus is produced at the hind end of the female and mating takes place. Very soon the production of eggs begins and continues until the female body contains little else; she then dies and her body wall forms a brown, leathery cyst which is lemon-shaped and about $\frac{1}{25}$ in. long. The immature eggs continue to develop inside the cyst until a coiled larval eelworm is formed in each, ready to hatch and emerge under suitable circumstances. The cysts remain in the soil after the host crop has been harvested and lie dormant until the next host crop is grown. An old cyst may be entirely empty but new ones often contain 200-300 or more eggs. In the interval between host crops, some eggs hatch every year and others are destroyed by enemies or disease. Thus the eelworm population declines but, after a heavy infestation, some encysted eggs can still be found in the soil after as many as ten years.

The activity and rate of growth of the eelworms vary with soil temperature. In the spring when soil temperatures are low, the life cycle is completed in about ten weeks, whereas in summer when soil temperatures are higher, only six weeks are required. As soil temperatures fall in the autumn, growth slows down and almost ceases in October and November. Young eelworms entering plant rootlets after the end of August fail to complete their development. The number of generations in a year depends on the growing period of the host crop: on the sugar beetroot crop there is normally time for two to two-and-a-half generations before lifting; on summer rape, mustard and brassicas which are in the ground from March to July there is only time for one generation. Normally, conditions which favour the host crop also favour the eelworms.

CROP SYMPTOMS AND HOST PLANTS

Injury to sugar beet or mangel plants is caused by eelworms feeding within the rootlets. The eelworms are so small, however, that a few have no serious effect on growth or yield and they pass unnoticed. Only when numbers are very great because of too frequent cropping does injury become apparent as patches of stunted plants (Table 1 and Plate XI). Fields in which there are no obvious symptoms of eelworm stunting and where there are few cysts in soil samples or on roots may, nevertheless, harbour very large numbers. The absence of cysts from a soil sample as large as 2 lb does not mean that the field

## TABLE I

*Crop symptoms and beet cyst eelworm numbers in soil*

| Crop symptoms | Cysts with living contents per 2 lb (kilo) of soil | Equivalent number of eggs (assuming that a cyst with contents contains an average of 100 eggs) | |
|---|---|---|---|
| | | per g soil | per acre |
| Beet-sick . . . . . | 1,000 | 100 | 1,000 million |
| Sick patches . . . . . | 100 | 10 | 100 million |
| No apparent symptoms . . . $\left\{\begin{array}{c} \\ \\ \\ \end{array}\right.$ | 10 | 1 | 10 million |
| | 1 | < 1 | 1 million |
| | 0 | — | 0–1 million |

is uninfested: there may be up to a million eggs per acre. In the worst affected patches tap roots are absent, lateral roots have proliferated excessively and many are dead, so that the plants are poorly anchored in the soil (Fig. 4(*d*) and Plate VIII). Tops tend to wilt in sunshine, or on dull days if the soil is dry. Sick patches (Plate XI) become apparent early in June. Hot weather and especially drought in June and July make the patches more obvious and probably increase yield losses. Similar patches of stunted plants may arise from other causes, but eelworm-infested plants can usually be distinguished by the white females on their rootlets (Plate VIII). Although there may be a recovery in the size of tops in the autumn with the development of a second growth of lateral rootlets, the main root remains small and stunted. By the time sick patches appear, beet cyst eelworm has been present for years and has spread to other parts of the field and to neighbouring fields. Sick patches often appear first near gateways or on headlands where beet has been piled, and are a warning that sugar beet, red beet, mangels or cruciferous host crops are being grown too frequently somewhere on the farm. If the warning is not heeded, the whole field may become sick so that profitable crops of beet or mangels cannot be grown again for many years.

Many crops and weeds are hosts of beet cyst eelworm, the most important being:

### Crop Plants

| Beet Family | Dock Family | Cabbage Family | |
|---|---|---|---|
| Sugar beet | Rhubarb | Cabbage (all kinds) | Rape or Coleseed |
| Mangel | | Kale (all kinds) | Kohl-rabi |
| Fodder Beet | | Brussels sprout | White mustard |
| Red Beet | | Broccoli | Black mustard |
| Spinach | | Cauliflower | Trowse mustard |
| Good King Henry | | Turnip | *Radish |
| | | Swede | Cress |

### Weeds

Beet Family
   *Fat hen (*Chenopodium album* L.) and other *Chenopodium* spp.

Cabbage Family
   Charlock (*Sinapis arvensis* L.)
   Shepherd's purse (*Capsella bursa-pastoris* (L.) Medic.)
   *Wild radish (*Raphanus raphanistrum* L.)
   Field penny-cress (*Thlaspi arvense* L.)
   Treacle mustard (*Erysimum cheiranthoides* L.)

---

* Poor hosts on which few cysts form.

Dock Family
  Broad-leaved dock (*Rumex obtusifolius* L.)
  Curled dock (*Rumex crispus* L.)
  *Persicaria (*Polygonum persicaria* L.)
Chickweed Family
  *Chickweed (*Stellaria media* (L.) Vill.)
Dead Nettle Family
  *Large-flowered hemp-nettle (*Galeopsis speciosa* Mill.)

* Poor hosts on which few cysts form.

The wild beet *Beta vulgaris* L. ssp. *maritima* (L.) Thell. has no greater resistance to eelworms than commercial varieties of sugar beet, all of which are about equally susceptible. Sugar beet is not attacked by the potato cyst eelworm (potato root eelworm, *Heterodera rostochiensis* Woll.), which is now very widespread. The cabbage cyst eelworm (cabbage root eelworm, *Heterodera cruciferae* Franklin) attacks plants in the cabbage family and may be confused with the beet cyst eelworm but does not attack sugar beet.

### EFFECT OF CROPPING ON EELWORM NUMBERS

When a host crop is grown, the number of eelworms in the soil usually increases but may sometimes decrease. The rate of increase is greatly affected by the number present when the crop is sown: when few, the root system is healthy and many females develop, but when many, the roots are damaged and fewer are able to develop. Increase rates as great as 1,640 times have been observed experimentally in pots (over two generations) and rates exceeding 50 times in small plots. The average rate in beet grown in a four-year rotation is usually about ten times. The rate of increase on different host plants is important practically. Cruciferous seed crops, such as black mustard, white mustard, turnip seed and cress, increase numbers faster than sugar beet, brassicas increase numbers at about the same rate or a little faster, whereas spinach, chickweed, fodder radish and fat hen allow numbers to decrease at a rate equal to or greater than the decline when land is fallow. Swine-cress (*Coronopus squamatus* (Forsk.) Aschers.) and *Beta patellaris* are classed as enemy plants because they produce root diffusates that stimulate eggs to hatch, but the larvae which emerge and enter the roots are unable to mature. Although some weeds may be better hosts than those tested in experiments, weeds do not cause an increase in eelworm numbers unless the land is very weedy; they do, however, help the infestation to persist longer.

Catch crops of mustard, coleseed or turnips planted after early potatoes, early carrots or peas cause a smaller increase in eelworm numbers than do full crops. When sown in July, if germination is prompt, white females appear on the roots in four weeks and the first eggs are produced within six weeks of sowing. Therefore, to prevent an increase in the eelworm population such crops should be ploughed in within six weeks of sowing, which is too short a growing period to be worth while. When a catch crop is required for green manure, grazing, cover against wheat bulb fly or for game, fodder radish is the most suitable cruciferous crop. It grows vigorously, is leafier and yields more heavily than white mustard. Fodder radish is not a trap crop and does not actively decrease populations of beet cyst eelworm. When grown between July and October or November, it allows beet cyst eelworm populations to decline to about the same extent as a fallow. In the British Sugar Corporation's contract with growers (Clause 13) and in the administration of the Beet Eelworm Orders (see page 78), fodder radish is now classed as a non-host crop and may therefore be grown in any year preceding beet. In contrast, a catch crop of white mustard or other brassica, even if only grown for a few weeks, counts as a host crop in that year of the rotation and beet cannot be grown for the next two years, or longer if the land is known to be infested with beet cyst eelworm.

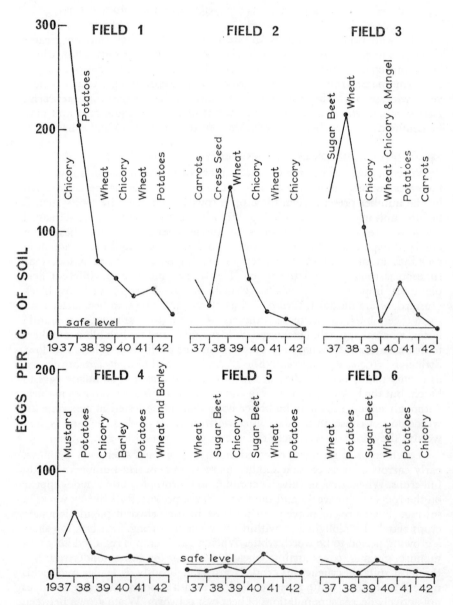

Fig. 13. Changes in the numbers of beet cyst eelworm eggs in soil under different crops. Fields 1–3 were extensively and heavily infested; fields 4–6 were less heavily infested and the infestations were mainly in a few patches.

When non-host crops are cultivated or the land is left fallow, the eelworm population falls during the year to about half of its initial level; this rate of decrease continues steadily every year that a non-host crop is grown and is largely independent of population density. As a result, moderate populations reach a safe level within three or four years but large populations may take much longer.

Fig. 13 shows examples of variation in soil populations in fields studied for several years. Fields 1, 2 and 3 were beet-sick and the numbers of eelworm eggs in the soil in each were enormous. In the most heavily infested part of field 1, there were more than 100,000 eggs per lb of soil. In fields 1 and 2, mangels were grown frequently before 1921 and either mangels, beet or cress seed were grown in eight of the next sixteen years. After 1936, sugar beet growing was abandoned in both fields, but in field 2, cress seed was grown in 1938 and caused an increase in eelworm numbers. Field 1 required fully ten years' rest from host crops before it was again safe to grow sugar beet or mangels. Test plots planted in 1944 and 1945 produced beet with normal tap roots but, at the end of the season, side roots were rather numerous and carried an abundance of females.

Numbers in the less heavily infested fields (4, 5 and 6) increased when host crops were grown, but decreased again to a safe level within two or three years. In all three, the distribution of the eelworms was patchy; the larger rise with the second beet crop in field 5 was mainly due to an increase in the size of the patches.

Because the beet cyst eelworm is most damaging to sugar beet and its allies and much less so to cruciferous crops, farmers often wonder whether cruciferous crops seriously increase numbers. Field observations and experiments, and a wealth of experience in Germany, the Netherlands, Scandinavia and the U.S.A., underline only too clearly the perils of disregarding the danger of cruciferous host crops on farms where beet is grown.

INFLUENCE OF SOIL TYPE

The beet cyst eelworm occurs in soils ranging from light sand to clay, but crop losses are greatest in light soils. Soils are usually classified according to the proportions of coarse sand, fine sand, silt and clay which they contain, but suitability for eelworms is determined not so much by the size range of the soil particles as by the degree to which these are aggregated into crumbs. The eelworms inhabit the pores between the crumbs and their movements are restricted by pore necks which must be large enough for them to pass. Most soils in seedbed conditions are porous and suitable for beet cyst eelworm, and by the time some coarse sandy soils have been compacted by rainfall, the eelworm larvae have entered plant roots and the life cycle is complete.

A survey in Belgium suggested that beet cyst eelworm distribution is influenced by soil pH. Although the eelworm was present in fields of pH less than 6·5 (i.e., acid soils), few were moderately or heavily infested, but this may reflect the relative infrequency with which beet is grown, or the poor growth of beet, on acid soils. With increasing pH (i.e., more alkaline soils), the proportion of fields in the heavily infested class also increased. The most alkaline soils were near beet sugar factories and had received large dressings of factory waste lime to improve their workability; probably these fields grew beet crops too frequently. Experiments in England suggest that increasing the concentration of potassium in soils decreases the number of female eelworms on the root systems of sugar beet plants.

SPREAD

Soil containing eelworm cysts is spread naturally by wind and water currents and on the feet of animals. However, these agencies are far less important than farm traffic (e.g., tractors, trailers, lorries), implements, boots, sacks, seed potatoes, bulbs, stecklings and other transplants. Cysts may also occur in any small lumps of soil among seeds, especially the large kinds, but as natural sugar beet seed is not now sown it is no longer a means by which the eelworm is spread. Much spread occurs at lifting time when soil adhering to the sugar beet gets scattered about on fields and road-ways during carting. All factories have washing guns and overhead roadways for tipping lorries, so beet is rarely forked from lorries and few leave factories with significant amounts of waste soil, i.e., enough to need tipping. Many thousands of tons of waste soil and trash accumulate annually at the beet factories and are deposited in semi-permanent dumps. Official recommenda-tions have been made for the further use of this soil and advice may be obtained from the Divisional offices of the Ministry of Agriculture, Fisheries and Food. Lorries and trailers carrying beet on or between farms sometimes contain small quantities of waste soil which has dropped off beet and this may be emptied by the driver regardless of the possibility of contaminating arable land. Spread within the farm and from farm to farm may also occur in much the same way during the harvesting of any other crop to which dirt normally adheres. Lime sludge from beet sugar factories cannot spread eelworm cysts as it is not contaminated with waste soil.

It is most undesirable to infest clean land with cysts, but introduction of small numbers is not serious as long as a sound rotation is followed. Substantial amounts of infested waste soil from beet piles and cleaner-loaders may be more serious as they may contain enough eelworms to cause small sick patches immediately. To minimize the spread of beet cyst eelworm from field to field within a farm, beet should be piled in the field where it is grown. If, for convenience, piles are made beside roadways or on a concrete platform, the waste soil that accumulates should be returned to the field from which the beet came, or spread evenly along the road verge or dumped in a pit. It is worth remembering that other soil pests and diseases and other cyst eelworms are spread in waste soil.

CONTROL

Much time and money have been spent trying to find chemicals to destroy cyst-forming eelworms in the soil. In the U.S.A. the soil fumigant dichloropropane-dichloropropene mixture, a volatile liquid which vaporizes when injected into the soil, is used as a supplement to crop rotation. The best results are obtained on light sandy soils; treatment is less effective on peats and clay-loams. Where eelworms are very numerous fumigation increases yield appreciably, but those that survive multiply on the healthy root systems and numbers are often greater after harvest than they were before sowing. Cyst eelworms are intrinsically difficult to kill; the quiescent larvae are protected by their own cuticles, their egg shells and the cyst walls, and the cysts themselves may be embedded in soil crumbs not easily penetrated by vapour. Free-living eelworms that cause Docking disorder move in the major passages in the soil and are protected by their own cuticle only; they are therefore more readily reached and killed by fumigant vapours (see page 88). Experiments are continuing with conventional soil fumigants like dichloropropane-dichloropropene mixture and with solids (e.g., dazomet) which break down when mixed with soil and liberate toxic vapours. Systemic nematicides, especially carbamoyl oximes, offer hope for the future.

Some wild species of beet (*Beta patellaris, B. webbiana* and *B. procumbens*) are resistant to beet cyst eelworm: scarcely any larvae that invade the roots succeed in developing into mature females. Plant breeders are attempting to incorporate this resistance into commercially acceptable varieties. In addition, beet strains tolerant to cyst eelworm damage have been developed, and some experimental varieties have already outyielded commercial varieties on heavily infested land and are being studied further.

It is far too hazardous to attempt to decrease beet cyst eelworm numbers by growing a host crop and destroying it when eelworms have been induced to hatch and have invaded the rootlets, for delay in ploughing-in may increase rather than decrease numbers. On the other hand, if the crop is destroyed soon enough, its roots do not have sufficient time to penetrate all parts of the soil, consequently a good hatch is not induced. The method cannot therefore be recommended.

Because at present there are no satisfactory chemical measures or resistant or tolerant varieties for use against the beet cyst eelworm, yield loss is minimized by crop rotations in which the interval between beet and other host crops is sufficient for eelworm numbers to decrease until little yield is lost. In Europe, where beet cyst eelworm has long been known, yields are satisfactory when beet or other host crops are grown only once in six years. On fertile soils of medium texture, a four-course rotation is usually possible but, when eelworms are numerous from too frequent cropping, some yield losses are experienced whenever beet is grown. Continental experience shows that serious losses occur after brassica crops are grown or a strict rotation is not followed. Beet on light soils usually suffers most from eelworm attack, possibly because such soils are porous, eggs hatch rapidly in them and larvae can move freely through them, or because their poor water-holding capacity makes plants with impaired root systems less able to obtain moisture and nutrients. If light soils become beet-sick it is often difficult to re-establish a four-course rotation even after resting the land for a long time. Numbers increase more slowly in silt soils than in peats but also decline more slowly. The silt soils of Lincolnshire are similar to the silt soils of the Netherlands and contain almost the same proportions of sand, silt and clay but less organic matter and a smaller percentage of calcium carbonate. As a result of overcropping with sugar beet and cruciferous oil-seed crops during World War II, silt soils in the Netherlands became beet-sick and it took many years for numbers to decline to acceptable levels. Overcropping with beet and cruciferous crops on the Lincolnshire silts might have similar consequences.

Probably there is no one safe rotation for any class of land. The increase in eelworm numbers and their effect on the crop varies with soil, season and host crop. As beet cyst eelworm becomes more widespread it may be necessary to sample soils, count the number of eelworms present and give advice accordingly, as is done in the Netherlands.

Crop rotation is enforced through the rotation clause in contracts with the sugar factories and also, in infested areas, through the Beet Eelworm Orders, 1960 and 1962. The British Sugar Corporation's contract with the grower (Clause 13—Crop Rotation) permits beet to be grown only after land has been free from the host crops (see page 72) for two years. The only exceptions are: (1) radish (including fodder radish) is excluded from the list of host crops, and (2) beet may be grown two years in succession or two years in three immediately after a ley of at least three years' duration. The Beet Eelworm Orders restrict the growing of host crops on known infested fields wherever they may occur and on uninfested fields in certain specified or scheduled areas where beet cyst eelworm is prevalent; the scheduled areas are parts of the Fenlands and a few infested sewage farms in beet-growing areas (Figs. 14 and 15). The Orders also stipulate that seed

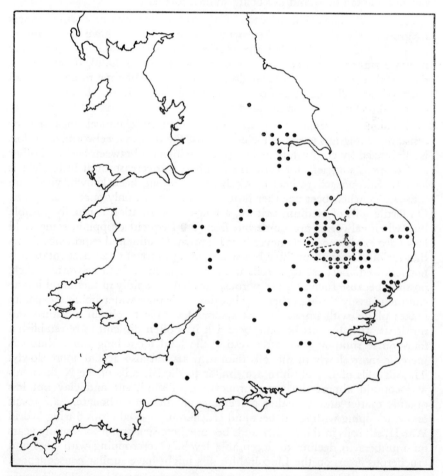

Fig. 14. Known distribution of beet cyst eelworm in 1970. Each spot represents at least one record in a 10 Km. square of the National Grid. Broken lines indicate areas scheduled under the Beet Eelworm Order, 1960.

potatoes and plants raised on infested land shall not be sold for planting or planted elsewhere than on the farm or holding on which they were grown. On infested fields, sugar beet and other crops can be grown only under licence from the Minister of Agriculture, Fisheries and Food or the County Agricultural Executive Committee. The terms of the licence are generally as follows:

> *Infested fields*—a host crop may be grown only after the land has been free from host crops for *three years*.

> *Heavily infested fields*—a host crop may be grown only after the land has been free from host crops for *five years* and a soil sample shows that the eelworm population has decreased sufficiently. Thereafter, the land shall be treated as 'infested'.

On uninfested fields within the scheduled areas, a licence is also required for departure from a three-course rotation, i.e., a host crop may be

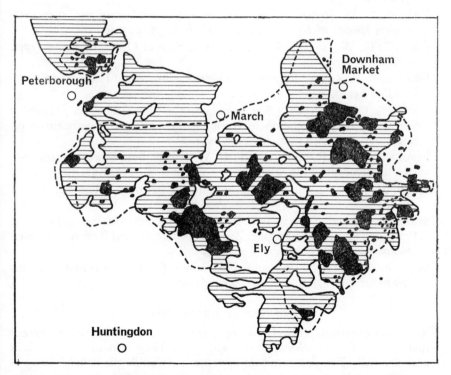

Fig. 15. Known distribution of beet cyst eelworm in the Fens in 1964. The black areas are infested, shaded areas are fen peat soils and the dotted line marks the boundary of the areas scheduled under the Beet Eelworm Order, 1960

grown only after the land has been free from host crops for *two years*. For the purposes of the Orders, the usual unit is a field because its boundaries offer the first real resistance to further spread. When part of a field is infested, experience shows that the whole soon becomes infested, even if the land is occupied by several tenants. Strip cropping makes control more difficult because within a few years the exact location of a crop is easily forgotten.

The field staffs of beet sugar factories keep a look out for new infestations and survey annually a sample of fields outside the areas where beet cyst eelworm is known to occur. At present few fields are very heavily infested (i.e., beet-sick), although many are lightly infested and more are found every year. Growers who find eelworm cysts on the roots of beet, mangels or red beet should inform the British Sugar Corporation or the nearest Divisional Office of the Ministry of Agriculture, Fisheries and Food, even though the growth of the crop may not apparently be affected.

FURTHER READING

CURTIS, G. J. (1964). The effect of potassium chloride on the infestation of sugar beet by beet eelworm, *Heterodera schachtii* Schmidt. *Ann. appl. Biol.* **54,** 269–80.

JONES, F. G. W. (1950). Observations on the beet eelworm and other cyst-forming species of *Heterodera*. *Ann. appl. Biol.* **37,** 407–40.

JONES, F. G. W. (1951). The Sugar Beet Eelworm Order 1943. *Ann. appl. Biol.* **38,** 535–7.

JONES, F. G. W. (1955). Report on a visit to Germany and Holland. *Brit. Sug. Beet Rev.* **24,** 25–8, 39 and 77–9.

JONES, F. G. W. (1957). Soil populations of beet eelworm (*Heterodera schachtii* Schm.) in relation to cropping. III. Further experiments with microplots and with pots. *Nematologica* **2,** 257–72.

PETHERBRIDGE, F. R. and JONES, F. G. W. (1944). Beet eelworm (*Heterodera schachtii* Schm.) in East Anglia 1934–1943. *Ann. appl. Biol.* **31,** 320–32.

SIMON, M. (1954). L'étude du rapport entre le pH du sol et les nématodes. *Publs. tech. Inst. belge Amélior. Better.* **22,** 85–99.

STEUDEL, W. and THIELEMANN, R. (1970). Weitere Untersuchungen zur Frage der Empfindlichkeit von Zuckerrüben gegen den Rübennematoden (*Heterodera schachtii* Schmidt). *Zucker* **23,** 106–9.

WINSLOW, R. D. (1954). Provisional lists of host plants of some root eelworms (*Heterodera* spp.). *Ann. appl. Biol.* **41,** 591–605.

ROOT-KNOT EELWORMS*

Root-knot eelworms (*Meloidogyne* spp.) are similar to cyst eelworms except that the swollen females do not form cysts. They remain embedded in knot-like galls (Fig. 4(*d*)) and the eggs are laid outside the female's body in a gelatinous egg sac. They are serious pests of sugar beet in warmer countries such as Italy and parts of the U.S.A. Occasional attacks are reported in Britain. Sometimes they occur where infested glasshouse soils have been dumped on agricultural land; these infestations die out because the species usually found in glasshouses on tomatoes and cucumbers probably originated

---

* See also Advisory Leaflet 307, available from the Ministry (p. 108).

in warmer countries and are not well adapted to outdoor life in Britain. Natural field infestations of another species, the northern root-knot eelworm (*Meloidogyne hapla* Chitwood), which survives outdoors in our climate, occur on light sandy soils in East Anglia.

Another root-knot eelworm (*Meloidogyne naasi* Franklin) occurs in Wales and the southern and western counties of England from Hampshire and Cornwall to Lancashire. It also occurs on the Continent and in the U.S.A. In Belgium it severely stunts beet on light sandy soil. Although in Britain it attacks sugar beet, it is usually found on barley, wheat and fodder grasses. On cereals, small galls are formed without much proliferation of lateral rootlets; the galls are club-shaped when terminal, but when behind the root tip they curl the infested rootlet into a horse-shoe shape or a loose spiral. On beet, the galls are mainly on lateral roots close to the tap root (Plate VIII). Galls caused by the northern root-knot eelworm on beet are similar but are associated with more root proliferation. Little is known about the damage these two species of eelworm cause to beet in England and Wales and no recommendations can be made for control.

FURTHER READING

BROWN, E. B. (1955). Occurrence of the root-knot eelworm, *Meloidogyne hapla*, out of doors in Great Britain. *Nature, Lond.* **175,** 430–1.

FRANKLIN, M. T., CLARK, S. A. and COURSE, J. A. (1971). Population changes and development of *Meloidogyne naasi* in the field. *Nematologica* **17,** 575–90

D'HERDE, J. (1965). Een nieuw Wortelknobbelaaltje, parasiet van de Bieteteelt. *Meded. LandbHoogesch. OpzoekStns Gent* **30,** 1429-36.

STUBBY-ROOT EELWORMS AND NEEDLE EELWORMS
(DOCKING DISORDER)

Docking disorder is characterized by irregularly stunted plants, and often by fangy root growth, on light, structureless, sandy soils containing little clay or organic matter. Although many factors seem to be involved, the poor growth of the beet is caused primarily by stubby-root eelworms (*Trichodorus* spp.) and needle eelworms (*Longidorus* spp.) (see page 66; they feed on the developing root tips and impair their function or kill them, preventing the roots from obtaining enough nutrients, especially nitrogen and magnesium. The eelworms begin to feed on the young roots of seedlings immediately after germination and continue to do so throughout the season, sometimes with a break in July-August if there is drought.

The disorder is named after the parish of Docking in west Norfolk where, as an unexplained trouble, it was studied in 1948 and subsequently. Docking disorder almost certainly occurred before 1948 but was confused with the effects of acidity, poor drainage, soil compaction, variations in soil fertility and structure, and other causes of patchy growth. Recent research has shown that, at Docking, it is caused mainly by stubby-root eelworms. In other fields in East Anglia, Lincolnshire and, especially, east Yorkshire, it seems also to be caused mainly by stubby-root eelworms, whereas needle eelworms are more important in some fields in Norfolk, Suffolk and the West Midlands (Fig. 16). As the cause of Docking disorder was unknown for many years, the name became loosely used for other types of poor growth of unknown cause; those not caused primarily by eelworms and therefore not considered in this Bulletin to be Docking disorder are described briefly on page 88.

FIELD SYMPTOMS

When the seedlings emerge they are uniform in size, but as early as the two rough-leaf stage differences in growth rate become apparent in diffuse patches. These patches of affected plants are irregular in shape and extent, the variation in plant size increases and the worst affected patches appear gappy. The haphazard intermingling of large and stunted plants (the 'hens' and 'chicks') is very characteristic (Plate X). Stunted plants sometimes occur singly, sometimes in groups, patches or in lengths of row.

The 'hen' plants grow normally and appear healthy, except perhaps for slight magnesium deficiency; sometimes they are extremely vigorous and become ten or twenty times bigger than the 'chicks'. The stunted plants remain small and the leaves often show symptoms of nitrogen, magnesium and other deficiencies (Plate X).

Usually the affected plants, the 'chicks', recover to some extent later in the year. When recovery is rapid, yield is not seriously affected, but when it is slow and not until September, root yield may be as little as three or four tons per acre, the stunted plants being unharvestable and the yield derived only from the scattered large ones. In one field studied in detail there were in the affected area about 20,000 plants per acre yielding four tons of washed beet containing 15 per cent sugar, whereas in the unaffected area there were 25,000 plants per acre yielding nineteen tons of beet with 17·2 per cent sugar. A needle eelworm, *Longidorus attenuatus* Hooper, was abundant in the deeper and coarser soil in the affected area but absent in the shallower and finer-textured soil in the good area.

Whether the damage in a particular field is from stubby-root or needle eelworms cannot be determined by leaf symptoms or by the distribution of affected plants. Root symptoms give some indication which eelworm is mainly responsible (Plates IX, X and Fig. 4(d)), but this can be confirmed only by sampling the soil, extracting, identifying and counting the eelworms present.

ROOT SYMPTOMS CAUSED BY STUBBY-ROOT EELWORMS

Stubby-root eelworms have short stylets able only to pierce and extract sap from the outer layer of root cells. They are attracted by actively growing roots and congregate behind the growing tips where their feeding causes browning and cracking. Severely attacked seedling root systems cease to grow and the rootlets become stubby-ended, turning grey-brown and later black as they die and decay (Plate IX). New roots are put out near the soil surface and may escape heavy attack, possibly because the soil is drier there and less favourable for eelworm movement. The crippled tap root may die at an early stage or linger on as a useless relic. Other rootlets replace its function and grow horizontally or diagonally but rarely straight down, thickening and so producing a fangy root (Fig. 4(d)). Exceptionally, there may be as many as 5,000 stubby-root eelworms per pint (9,000 per litre) of soil around stunted plants; usually there are fewer but still many times more than around vigorous plants in affected patches and very many more than around plants in areas in the same field where growth is normal.

Several species of stubby-root eelworm are associated with Docking disorder. Fields in the Docking area of Norfolk contain mixtures of *Trichodorus cylindricus* Hooper, *T. primitivus* (de Man), *T. pachydermus* Seinhorst, *T. viruliferus* Hooper and *T. teres* Hooper in varying proportions. In Yorkshire *T. teres* is common in some fields and *T. anemones* Loof in others. In the Netherlands *T. teres* (reported there as *T. flevensis* Kuiper and Loof) stunts beet and causes fangy roots in polder soils. In many light sandy soils stubby-root and needle eelworms occur together; usually one kind of eelworm is more numerous than the other and they are rarely numerous together. *Trichodorus* spp. have a shorter life cycle than have *Longidorus* spp. and injury may therefore be increased by a rise in numbers during the season. While stunting by stubby-root eelworms is more widespread and more serious nationally than stunting by needle eelworms, individually the latter are larger and more injurious, and fewer per seedling can retard growth.

## ROOT SYMPTOMS CAUSED BY NEEDLE EELWORMS

Stunted plants from patches where there are many needle eelworms usually have long tap roots but many short and poorly developed lateral roots (Plate X and Fig. 4(d)). Needle eelworms have long, slender stylets which they insert into the growing points of root tips; few of the lateral rootlets of attacked plants develop properly and most have dead or dying tips with swellings immediately behind the tip (Plate IX). Occasionally a lateral root that has grown normally may thicken and extend abnormally in a horizontal direction just below the soil surface.

*Longidorus attenuatus* is the eelworm species associated with injury of this type in sandy soils in East Anglia. *L. elongatus* (de Man) is injurious in coarse sandy soils elsewhere and is also injurious to cocksfoot, meadow fescue, carrots and kale in some light fen peat soils in Norfolk and Suffolk. The symptoms in these crops are much the same as those in beet. Where sugar beet is stunted there are up to 1,200 needle eelworms per pint (2,000 per litre) of soil around poor plants but many fewer around more vigorous plants. When numerous, needle eelworms can be seen coiled among rootlets and between soil aggregates.

In a few fields a sheath eelworm, *Hemicycliophora similis* Thorne, occurs along with stubby-root and needle eelworms. Like needle eelworms it has a long stylet and feeds on root tips. When roots are lifted, the eelworms remain attached to the tips by their stylets. Root injury resembles that by needle eelworms, but whether sheath eelworms stunt beet seedlings is unknown.

## FACTORS AFFECTING THE INCIDENCE OF DOCKING DISORDER

The circumstances that cause annual variations in the severity of Docking disorder are imperfectly understood and the subject of much research. The most important factors affecting incidence are the number and activity of eelworms in the soil at the time of seed germination; these are influenced especially by previous cropping, soil structure and moisture. Plants are more susceptible to injury if they are growing poorly because of unfavourable soil conditions, too deep drilling, herbicide toxicity, lack of adequate nutrients, or other adverse factors.

When the seedling puts down its first root, eelworms soon find it; if many eelworms attack it, proper root development is impossible, the seedling plant fails to obtain nutrients from the soil and so cannot grow. The early stages of root growth are probably the most critical and some means of protecting seedlings from eelworm damage is obviously needed; ways of providing this are being studied (see page 88).

*Rotation.* Although grass leys improve fertility, they seem also to favour needle eelworms so that beet grown after grass is sometimes severely attacked. On light land, sugar beet normally follows a cereal, usually barley. If eelworms

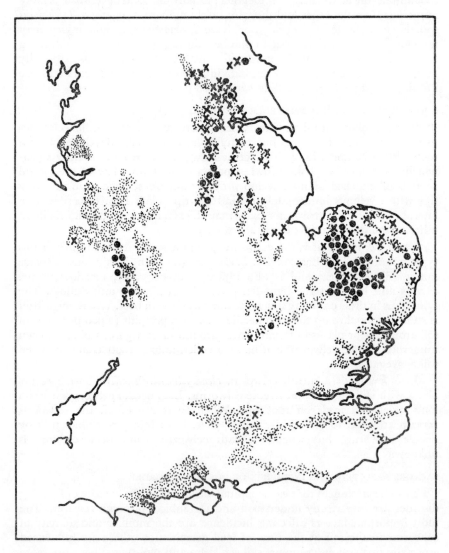

Fig. 16. Map of the principal areas of sandy soil in England showing the distribution of fields where sugar beet has been stunted up to 1971.

× — stunting mainly by stubby-root eelworms;

● — stunting mainly by needle eelworms.

Each symbol represents one or more fields. Needle eelworms are scarce in the north-east, whereas in other areas both kinds usually occur together.

are present in the soil they are often numerous after barley crops, so that the apparent increase in the incidence of Docking disorder over the last ten years may be partly because of the increase in barley growing. Barley is attacked by eelworms of this kind but there is little evidence that yield suffers. In Yorkshire, cereals are attacked by stubby-root eelworms, but winter wheat suffers most, possibly because soil conditions in early autumn, late winter and early spring enable the eelworms to feed longer than is possible in spring-sown crops. On soils where sugar beet is stunted, suitable crops are few and rotations are therefore limited; rotation seems unlikely to control the eelworms, which feed indiscriminately on most crops and weeds.

*Soil.* Where Docking disorder occurs in west Norfolk and west Suffolk the subsoil is usually chalky and the top soil commonly contains pieces of flint and chalk. Elsewhere the subsoil is sand, often over clay, and the top soil may be free of stone. Top soil rarely exceeds 9–12 in. in any area and deep ploughing brings up chalky subsoil or raw sand. The soils are usually alkaline from mixture with subsoil, or from marling or liming, but are occasionally neutral or even acid. The map in Fig. 16 shows the main sand land areas and the sites where Docking disorder caused by stubby-root eelworms or needle eelworms has been found. Docking disorder is confined to areas of sandy soil on which about 70,000 acres of sugar beet are grown annually. Similar troubles occur in similar soils elsewhere in Europe and the U.S.A.

Coarse soils provide the most suitable environment for the eelworms. Soil variations influence the activity of eelworms as well as plant growth; it is not therefore surprising that the patchiness so characteristic of Docking disorder is often linked with changes in soil structure or texture. Where areas of badly stunted plants are close to areas of large plants, the latter are usually on soil containing more clay. Within poor patches, although the top soil around large plants usually differs little from that around stunted plants, the subsoil may be very variable. The great differences in the depth and composition of the soil over short distances were caused by conditions in the ice ages. The present crop variations arising from these differences often form irregular polygonal patterns on level ground or long stripes down sloping ground; these can be seen in aerial photographs (Plate XI). Man-made disturbances of the soil, such as the sites of old roads or settlements, also cause patches of uneven growth. Beet often grows poorly on the edges of marl pits where the top soil is shallow and structureless. These effects should not be considered as Docking disorder unless the plants show the characteristic symptoms and eelworms are numerous in the soil around stunted plants.

Sandy soils where Docking disorder occurs are ploughed late and cultivated little to avoid over-fine seedbeds with the risk of capping or blowing. Rolling accentuates blowing and therefore beet may often be drilled too deeply into loose seedbeds, resulting in weakly plants which succumb readily to eelworm attack. Better growth is sometimes observed where the soil has been consolidated by tractor wheels during seedbed preparation; this may limit eelworm movement and help the plants to grow.

*Plant nutrients.* Stubby-root and needle eelworms so stunt seedling root systems that the plants are unable to obtain adequate nutrients from the soil; stunted plants are especially deficient in nitrogen and magnesium. The harmful effects of the eelworms are probably accentuated by the shortage of mineral reserves in sandy soils and the ease with which some nutrients, especially nitrogen and magnesium, are leached. The apparent infertility of the 'poor' sands on which Docking disorder occurs may be largely a combination of these two factors. Plants inadequately nourished grow poorly and are more readily stunted by eelworms feeding on their roots; a few eelworms may cause as much damage as would many eelworms in more fertile soil.

Crops severely stunted in May and June by needle eelworms sometimes recover in July and yield reasonably, especially where fertilizer has been liberally applied before drilling or later as a top-dressing. However, when stunting is by stubby-root eelworms, recovery in the growth of tops can be misleading for roots remain fangy and usually yield poorly. Eelworm movement and feeding are much influenced by soil moisture, and the cessation of feeding during a relatively dry period may possibly allow plants to produce new roots and grow away. Stunted plants washed free from soil and eelworms, and kept moist, rapidly put out new roots, showing that they have the capacity to recover if allowed.

*Season and soil moisture.* The acreage of beet affected by Docking disorder varies from year to year; Docking disorder was most prevalent in 1948, 1949, 1953, 1954, 1958, 1959, 1963, 1964, 1965, 1967 and, especially, 1969. Weather in the preceding year may influence the number of eelworms left by the previous crops, while weather in the current year influences eelworm activity and crop growth. For eelworms to be fully active, the force binding water to soil particles must be weak (no more than 100–150 cm of water suction), so the best conditions for them are when the soil is draining after rain. The first prolonged dry spell of the season probably creates moisture tensions great enough to stop eelworms moving and feeding but not nearly enough to inhibit root growth. During this period the roots probably proliferate and grow vigorously, as they do in moist chambers after washing them free of soil and eelworms. When rain falls again the eelworms resume feeding, but the ratio of root tips to eelworms has increased, nutrients have been taken up and the plants have established effective root systems. The pattern of rainfall and the incidence of dry spells appear to influence greatly the injury that ectoparasitic eelworms can do. In 1967 and 1969 the frequent periods of rainfall throughout May probably enabled eelworms to be active from planting until the first really dry spell in June—longer than in other recent years.

*Herbicides.* Band-sprayed, pre-emergence herbicides sometimes appear to accentuate Docking disorder. Rates of application suitable for the parts of fields with heavier soil may be excessive for the lighter parts, or fluctuations in pressure or in nozzle height can lead to over-dosing of lengths of row. Seedlings thus retarded almost certainly suffer more readily from injury by eelworms.

*Viruses.* Tobacco rattle virus (TRV) is transmitted by stubby-root eelworms and tomato black ring virus (TBRV) by needle eelworms. TRV causes 'spraing' disease of potatoes, which seriously affects the saleability of some varieties in light sand areas. Weeds are also hosts of both viruses and TBRV can be carried by many weed seeds. The viruses are transmitted to sugar beet when eelworms feed and usually remain localized in the roots, but sometimes spread throughout the plants and produce symptoms in the leaves. Leaves of plants infected with TRV may show bright yellow blotching and 'yellow blotch' is the name proposed for the disease, but the symptoms may be less obvious with an olive-green blotch or bands along the main veins. TBRV often produces a light and dark green leaf mottle not unlike that produced by beet mosaic virus, but TBRV does not distort the young leaves nor cause vein clearing. Infected leaves may show a ringspot symptom and the disease is called 'ringspot'. Plants are often infected with both viruses, but the symptoms tend to disappear or be obscured by symptoms of nutrient deficiency from mid-August onwards. Although these viruses may stunt beet plants, too few plants become infected to influence yield in most fields and many stunted plants are uninfected.

*Other pathogens.* Weakly parasitic fungi, especially *Rhizoctonia* spp., are often found on roots injured by stubby-root eelworms, and fangy roots can be produced by infecting plants experimentally with *R. solani*. The inter-relationships of eelworms and fungi are being studied. Apart from root-lesion eelworms (*Pratylenchus* spp.) and stunt eelworms (*Tylenchorhynchus* spp.), which are present in nearly all soils, no other pathogens have been found consistently in soil or roots where Docking disorder occurs.

CONTROL

In the fields where there is a history of Docking disorder or where eelworm counts suggest that trouble may arise, yield losses may be minimized by improving the soil structure, supplying adequate nitrogen, remedying deficiencies of minor elements and ensuring that the beet is sown under the best possible conditions; the more vigorously the seedlings grow, the less likely will they succumb to eelworm damage. If the soil is a light, blowing sand, marling should be considered and, where suitable marl is readily available nearby, the cost is not too great*. A leguminous crop before beet helps to improve soil fertility. In addition to inorganic fertilizers, an organic fertilizer such as farmyard manure or, especially, shoddy should be applied wherever possible, as these improve soil structure and release nutrients slowly. When organic fertilizers are not used, 120 units of nitrogen should be applied per acre as late as possible during seedbed preparation and more added as a top-dressing later if necessary. Kieserite applied at 3 cwt per acre in the autumn before sowing helps to make good any deficiency of magnesium. If boron deficiency seems likely, boron should be applied with the fertilizer for beet or later as a foliar spray. (Note that boronated fertilizer is toxic to cereals and must not be applied to them.) The seedbed should be firm, the crop sown neither very early nor very late, and too deep drilling must be avoided. If blowing occurs and the crop has to be re-drilled, the land should be cultivated as little as possible to retain firmness, and additional nitrogen should be added to the seedbed to replace that lost.

Little can be done to cure stunted crops in late May or June. The soil pH should be checked but no lime is needed if it exceeds 6·5. Top-dressing with a quick-acting nitrogenous fertilizer, at a rate depending on the amount of nitrogen applied to the seedbed and the rainfall since application, usually improves growth as soon as rain or irrigation washes the nitrogen into the root zone. When there is ample nitrogen in the soil already, more may improve

---

* Under the Farm Capital Grants Scheme, 30 per cent of the cost of an approved project is reimbursed by the Ministry.

top growth during the summer but not the yield. The effect of irrigation alone is unknown.

Soil fumigation with dichloropropane-dichloropropene mixture or dichloropropene can greatly improve the beet yield in badly affected fields. Complete treatment of the soil requires 200–400 lb (17–34 gal) per acre injected by special machinery to a depth of 6 in. in rows 10 in. apart. The liquid fumigant volatilizes slowly, kills most eelworms, some pathogens, weed seeds and insects, and mineralizes nitrogen locked up in organic matter. It also kills the bacteria that change ammonium nitrogen into nitrate nitrogen; whereas the latter is readily leached, the former remains in the surface layers. The bonus of readily available nitrogen that remains in the root zone may stimulate early and vigorous seedling growth. Where a complete treatment, i.e., a large quantity of nematicide, is applied, a long interval between treatment and sowing is essential otherwise the beet itself may be adversely affected by vapour remaining in the soil. Such treatment must be applied at least eight weeks before sowing and preferably before the end of December.

The cost of complete treatment is generally considered too great, but sometimes succeeding crops benefit from removal of their pathogens from the soil and the cost can be shared between the crops. The cost of soil fumigation may be decreased somewhat by combining treatment with winter ploughing, that is, dribbling the fumigant into the bottom of the plough furrow.

Fumigating the row positions protects the seedlings by killing the eelworms in the root zone or discouraging them from feeding during the most susceptible stage of plant growth. Apply dichloropropene, or its mixture with dichloropropane, 6–8 in. deep at 0·66 or 1 ml per ft of row respectively (4 or 5·6 gal per acre at 20 in. row width) and seal the coulter slit. The soil should be in seedbed condition when the fumigant is applied and must not be too wet.

Treatment at these rates shortly before drilling is usually, but not always, safe. An interval of 10–14 days or longer is advised when more than 6 gal of fumigant are applied per acre. Granular experimental nematicides, applied with the seed during drilling, control the eelworms well but none are yet available commercially.

The 'hen and chick' effect so characteristic of Docking disorder suggests that the potentialities of 'hen' plants as breeding material should be studied. Incorporation of resistance or tolerance to eelworms in commercial varieties of beet might provide a long-term answer to the problem.

STUNTING AND FANGINESS PREVIOUSLY CONFUSED WITH DOCKING DISORDER

Root stunting and fanginess are sometimes caused by acidity, water-logging, soil compaction, lightning strikes, excessive fertilizer, or pest damage; all are factors that can prevent normal root growth. None produces the almost horizontal root growth or the fanginess often characteristic of Docking disorder, particularly that caused by stubby-root eelworms, and especially characteristic of the severe patchy stunting known as 'Barney patch'.

Barney patch was first observed at Barney, Norfolk, in 1962 and at several other sites since. The bigger patches, which may be 25–100 yards across, are approximately kite-shaped with concave sides, the long axes of the kite being parallel to the principal directions of cultivation; there is a similarly shaped patch of normal beet in the centre. More commonly at the same sites and at others, there are one or more much smaller, scattered, roughly circular patches with no normal beet in the centre (Plate XI). In both types plants are severely stunted and may die leaving a patch with gaps in which weeds grow profusely. The tap roots of surviving plants are stumpy with swollen lateral roots (fangs) that run horizontally just below the soil surface (Plate X and Fig. 4(d)). Roots rarely penetrate more than 3–6 in. and most of the many fibrous roots that are produced soon die back from the tips and remain as a 'beard'. These symptoms are very similar to those produced by *Trichodorus* spp. (see page 82) but differ from them in that the patches often have a sharply defined edge where apparently normal and severely stunted plants grow almost side by side. Often the well-grown plants nearest the patch are greener and taller than normal plants in the crop, probably because they have no competition, so that the patch is outlined. The leaves of affected plants show no symptoms of virus disease and rarely any of nutrient deficiency, but they are sometimes bluish and wilt in dry weather.

These patches have been recognized in at least forty fields in recent years on sandy loam or silt soils in Norfolk, Suffolk, Lincolnshire and Yorkshire. Top-soil structure is virtually the same throughout the patch and the surrounding area, but the subsoil may vary considerably. The soil is fertile and patches most often occur in fields that have recently had grass leys. Barley and wheat may be affected in the year before or after beet, but in the former crops the edges of the patches are less sharp. The patches do not appear every year, which makes the problem difficult to study.

No pathogen seems specifically correlated with Barney patch and viruses have not been isolated from affected plants. At some sites the fungus *Cylindrocarpon* sp. is common on the roots of affected plants and, at most sites, chytrid fungi also occur: *Olpidium* sp. at most, *Polymyxa betae* Keskin at one site and *Cladochytrium caespitis* Griff. and Maubl. at another. There are many plant-feeding eelworms in the soil; at some sites root-lesion eelworms, *Pratylenchus* spp., are more prevalent in soil around stunted plants than in soil around normal plants. Until the cause is known, rational control methods cannot be attempted.

FURTHER READING

DRAYCOTT, A. P. and LAST, P. J. (1971). Some effects of partial sterilization on mineral nitrogen in a light soil. *J. Soil Sci.* **22,** 152–7.

DUNNING, R. A. and COOKE, D. A. (1967). Docking disorder. *Br. Sug. Beet Rev.* **36,** 23–9.

HEATHCOTE, G. D. (1965). Nematode-transmitted viruses of sugar beet in East Anglia, 1963 and 1964. *Pl. Path.* **14,** 154–7.

KUIPER, K. and LOOF, P. A. A. (1961). *Trichodorus flevensis* n. sp. (Nematoda: Enoplida), a plant nematode from new polder soil. *Versl. Meded. plziektenk. Dienst Wageningen* **136,** 193–200.

WHITEHEAD, A. G. (1965). Nematodes associated with 'Docking disorders' of sugar beet. *Br. Sug. Beet Rev.* **34,** 77–8, 83–4, 92.

165255

WHITEHEAD, A. G., DUNNING, R. A. and COOKE, D. A. (1971). Docking disorder and root ectoparasitic nematodes of sugar beet. *Rep. Rothamsted exp. Stn for 1970*, Pt. 2, 219–36.

WHITEHEAD, A. G. and TITE, D. J. (1967). Small doses of D-D soil fumigant to control free-living nematodes injurious to sugar beet. *Pl. Path.* **16,** 107–9.

# Bird and Mammal Pests

THE bird and mammal pests of sugar beet are familiar and need little description. They are active and very mobile throughout the year. Birds are especially mobile as their power of flight is far superior to that of insects, is more purposeful and is less influenced by air currents. Pigeons, rooks and other birds which form flocks may descend suddenly upon a crop and cause much damage in a short space of time before any action can be taken against them. Birds have an acute sense of sight and mammals an acute sense of smell; they both have complex patterns of behaviour and may modify them in the light of experience. Because of their highly developed senses they can be frightened off a field temporarily by a shout or a shot from the margin, which is more than can be said for any insect or allied pest! On the other hand, familiarity breeds contempt, so that deterrents are usually of only temporary value.

Birds have recently increased in importance as sugar beet pests. It is interesting to quote the introductory paragraph to the one page devoted to birds and mammals in the first (1957) edition of this Bulletin:

'Of the various birds and small mammals that may from time to time be found in the sugar beet crop only the rabbit and the rook are seriously troublesome. Hares sometimes feed on sugar beet but are rarely sufficiently numerous to do real harm, except perhaps in Scotland. Game birds peck the leaves of seedlings and domestic fowl attack plants near farm buildings.'

Damage by pigeons was not noticed by fieldmen until the late 1950s, small birds (mainly sparrows) fed only on beet near farm buildings, game bird damage to beet after singling did not occur on a noticeable scale until 1957, and in 1971 field mice first caused concern in many areas by taking seed from the soil. It is difficult to explain these changes except that seedling damage is now more noticeable because of wider seed-spacing. By contrast, rabbit damage has decreased with the decrease in rabbit numbers due to myxomatosis, and damage caused by rooks searching for wireworms has also decreased markedly.

Local or national feeling may hinder the control of warm-blooded pests, even when they are clearly harmful. Wholesale destruction and methods considered cruel soon meet with public objections, but little notice is taken of similar methods used against cold-blooded animals. Over a period of years a mass of legislation has accumulated which regulates the taking or killing of birds and mammals. The Pharmacy and Poisons Act, 1933, and the Poisons Rules, 1966, regulate the poisons, their handling and the animals against which they may be used. The use of poisons against mammals is governed by the Protection of Animals Act, 1911, the Protection of Animals (Scotland) Act, 1912, the Agriculture Act, 1947, the Agriculture (Scotland) Act, 1948, and the Animals (Cruel Poisons) Act, 1962, which restrict their use to rats, mice or other small ground vermin, permit the use of poisonous gases in holes or burrows to kill rabbits, foxes or moles and enable certain poisons to be prohibited by regulations. As the word 'vermin' in the Scottish Act of 1912 is not qualified by adjectives, poisons can be used legally against more animals in Scotland than in England. Under the Protection of Animals Act, 1927, it is a defence that poison was placed to kill insects, other invertebrates, rats, mice or other 'small ground vermin' in the interests of agriculture, provided precautions had been taken to protect dogs, cats, fowls, domestic animals and wild birds. The Prevention of Damage by Rabbits Act, 1939, made it legal to gas rabbits. The Pests Act, 1954, made gin traps illegal after 1958, but they may still be used against foxes and otters in Scotland until 1973. Some humane traps have been approved, but they must only be used in burrows or artificial tunnels, except to catch rats and mice when traps may be set in the open. The Protection of Birds Act, 1954 and 1967, and the many Orders made under it protect wild birds generally, but those listed or added to the Second Schedule of the Act, which include wood-pigeon, house-sparrow, rook and certain other corvids, may be killed or taken at any time by the owner, occupier or person authorized by them. There are no objections to funnel or cage traps, provided the traps are visited at least once a day and captured pests are killed humanely. Any control measures devised or used against warm-blooded pests must comply with these laws.

Specific recommendations for the control of bird pest damage are given below, but the problems and principles of prevention of damage by all of them are considered jointly on page 95.

The methods of feeding of the different bird and mammal pests are well known; the usual types of injury caused are given below, but the months during which feeding and injury occur are given on page 21.

Rook .. .. uproots young plants in searching for wireworms, etc.

Wood-pigeon .. may graze seedlings but causes most damage in June and July, stripping the leaves.

Pheasant .. grazes young seedlings and pecks into crown (below leaf level, above soil level) of plants after singling.

Partridge .. ⎫
Small birds .. ⎬ graze young seedlings.

Field mice .. dig up seed.

Rabbit .. .. ⎫ graze young seedlings and eat petioles, heart leaves
Hare .. .. ⎭ and crown of plants after singling.

Coypu .. .. bites off crown of young plants and eats into crown of older plants.

Some birds and mammals are rarely seen in the crop because they feed nocturnally or in the early morning. Others can often be seen in the crop, but binoculars and much patience are needed to observe them actually feeding on the beet plants. Damage has often to be ascribed to a particular species of bird or mammal on the symptoms of plant injury and its known presence in the area. Symptoms are not always clear-cut and characteristic, as feeding habits of birds and mammals vary much more than those of insects.

## ROOK*

Rooks cause losses by uprooting plants while searching for wireworms or other soil invertebrates. Often the plants have been singled and are growing

---

* See also Advisory Leaflet 244, available from the Ministry (p. 108).

Left: stunted seedling with many root-tip galls caused by *Longidorus* sp. Right: less heavily attacked seedling.

Left: healthy seedling. Right: seedlings stunted by stubby-root eelworms (*Trichodorus* spp.). Note darkening of tap root and lack of laterals except near soil level.

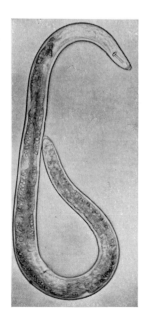

A typical plant-feeding eelworm. Note mouth stylet (x 150).

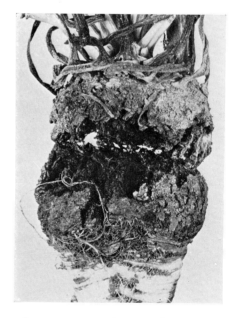

Severe crown canker caused by stem eelworm (x $\frac{1}{2}$).

PLATE IX

Docking disorder. Characteristic 'hen and chick' appearance, in July.

Root injury by needle eelworm in late July. Note symptoms of magnesium deficiency in leaves, also long, straight tap root (curled for photographing).

Plant from kite-shaped patch at Barney, west Norfolk, in July. No deficiency symptoms in foliage. Root is stunted with horizontal fangs.

PLATE X

Aerial photograph of beet field showing marked irregularity
in growth due to variation in soil and subsoil.

Patch of stunted and wilting beet in July caused by beet cyst
eelworm infestation.

Barney patch. Beet plants were stunted with very fangy roots
and many died. Warp soil near Goole in August.

PLATE XI

Small-bird damage; note beak mark on cotyledon of seedling
on the left.

Pheasant damage in June–July. Continued pecking at the
crown has felled the top.

Characteristic hare damage; some of the petioles are severed
and they and the growing points partly consumed.

PLATE XII

away. The intentions of the rooks may be admirable but the effect is disastrous, large bare patches being produced very quickly. The importance of the rook has declined in recent years, but local control of numbers may seem necessary in some years and the usual methods are shooting and trapping. Shooting young birds when they leave the nests has little significant effect on numbers since natural mortality at this time is considerable. Cage trapping can alleviate damage locally. A cage can be made by covering a light timber frame with 1½ in. mesh wire-netting; it should be approximately a six-foot cube with a door in one side for access. The top of the trap should be made to form a V-shaped trough 2–3 ft deep and having a 'ladder' type of entrance in the bottom with 6 in.-spacing between rungs. This type of trap has been found most successful. Grain, chaff or bread is placed in the cage as bait, also, if possible, trapped rooks, jackdaws or hens as decoys. Traps should be placed where rooks feed regularly and, if in fields, should be near a tree or fence where birds that are attracted may first perch. Rooks caught should be removed and destroyed after dusk each evening. (See also page 95 for control of bird damage).

## WOOD-PIGEON*

Originally the wood-pigeon or ring dove was a woodland bird feeding on tree buds or berries and on plants growing in glades and clearings. Its numbers have steadily increased with the development of modern agriculture, particularly intensive arable farming, which provides an abundant food supply, whilst wooded areas, hedgerows, copses and shelter belts form favourable nesting habitats. The wood-pigeon is now largely dependent on agricultural crops for its survival and is able to cause much damage to clover, brassicas and cereals.

Seedling beet crops are occasionally attacked in April and May, but pigeons cause most damage to beet in June and July when they feed on the leaves, sometimes leaving only the mid-ribs and main veins (Fig. 4(c)). Pigeon flocks may be initially drawn to a beet field by the presence of charlock, the leaves of which are attractive; they tend to alight where there are gaps in the stand or the plants are stunted, and they may continue to feed on the same field for several days. A large flock of pigeons can cause much damage locally in a short time. In summer and early autumn there is an abundance of other food, especially grain, and at this time the pigeons usually desert the beet fields.

Research has shown that population control of the wood-pigeon is not a practical possibility, so it is necessary to concentrate on crop protection where damage is actually occurring. This is probably best achieved by shooting, preferably over decoys, which has the double advantage of scaring some and removing others. Organized shooting, at dusk in winter, of birds returning to roost ('Battue shooting') is inefficient and is not recommended.

Under an appropriate licence it is also possible to use stupefying baits as a form of crop protection. Licences permitting the use of tic beans treated with alpha-chloralose can be obtained from Divisional Offices of the Ministry of Agriculture. To be successful, however, the method requires a considerable knowledge of fieldcraft and of local pigeon behaviour.

---

* See also Advisory Leaflet 165, available from the Ministry (p. 108).

## GAME BIRDS

Until 1957 relatively little damage to sugar beet by game birds was noted. Since then pheasant damage has increased and is pronounced in some years in well-wooded areas where game is 'keepered', especially in west Norfolk. Partridges and pheasants sometimes graze seedlings (see also page 12), but partridges are rarely sufficiently numerous in beet fields to cause concern. Pheasants can be particularly destructive after singling, when they peck into the crown of the beet plant just above soil level, eating the flesh and eventually felling the plant (Plate XII and Fig. 4(b)). This continues throughout the summer and autumn but is of less consequence after July as then the plants are too large to fell. Similar damage is caused by rabbits, rats and occasionally by hares, but their teeth marks are clearly visible, while the pheasant's beak leaves a much more pitted surface in a more hollowed cavity. All leave other tell-tale signs, such as small burrows, feathers or characteristic droppings. Plants are most often destroyed at the edges of a field, especially near shelter, or where the plants have gaps between them; thus the damage may appear worse than it is.

The cocks cause most damage, the hens being on the nest or with their chicks, and hard shooting of surplus cocks is recommended. In some areas cocks are even shot in the beet fields in May and June. This is illegal since the close season for pheasants is 2nd February to 30th September inclusive. Shooting rights make the problem more difficult and control by shooting may be impossible. Sometimes rent relief or compensation can be negotiated; sometimes the grower has little redress or, if a sportsman, does not seek it. Providing the pheasants with water or alternative food has been tried; supplies have to be placed on the margins of the beet field so that the pheasants can readily find them, but there is a risk of attracting pheasants from other fields. As damage still occurs in wet periods, the pheasants are not eating the beet for its moisture only. Mangels as an alternative food should not be spread around the fields unless entirely free of sprouts and aphids, otherwise there is a risk of introducing yellows viruses (see page 57). Aiming to grow 30,000–35,000 plants per acre is by far the best means of minimizing damage, and earthing up the plants slightly seems beneficial. (See also page 95 for control of bird damage.)

## SKYLARK AND OTHER BIRDS

Many birds feed on beet before singling; at first cotyledons are pecked off and sometimes also the growing point (Plate XII and Fig. 4(b)), but in the early rough-leaf stages the interveinal areas only are eaten, leaving a tattered or even skeletonized plant. Once the plants have passed the four-leaf stage, incidence of damage to the foliage by birds other than pigeons dwindles rapidly (see page 93). Defoliation severely retards plant growth; if the growing point is removed, leaving only the stem, the seedling dies.

The damage illustrated in Plate XII is characteristic of bird damage to seedling beet and cannot be caused by any small insect; large insect larvae such as cutworms and leatherjackets may eat the cotyledons and leave only stumps, but these insects are readily found in the soil. Bird damage can be confused with slug damage. Slugs remove areas of leaf in a rather similar manner to that of birds, and because neither pest can be found on the plant one cannot be sure which has caused the damage (see page 34). Rabbits and hares graze on young seedlings (see pages 96, 97) but their presence is usually known. A clean, straight cut is characteristic of bird damage to seedling; occasionally a V-shaped beak mark is left on the cotyledon, as shown in Plate XII. The cotyledons are eaten and only rarely left lying on the ground.

Some of the birds responsible for seedling damage in April and May have been discussed already (partridge, pheasant and pigeon), but other smaller birds are probably more important. Flocks of the common house-sparrow stunt or destroy plants in fields near farms and villages (see Advisory Leaflet 169). Skylarks are abundant in open fields and feed on seedling beet; such damage was reported on mangels in the 19th century when larks were said to take the cotyledons to their young. They occasionally also take any uncovered seed. Numbers can reach one breeding pair per acre, each with its own territory. Other common birds such as the moorhen, lapwing, yellow-hammer (bunting), linnet, greenfinch, chaffinch and pied wagtail may feed on young beet occasionally.

When heavy rates of natural seed were sown and pre-singling plant populations were far greater than they are now, this type of damage was of little consequence. Today, wider seed-spacing and the rapid change to mono-germ seed make it imperative that nearly all seedlings should survive. Injury by small birds, and indeed by other pests previously considered unimportant, is now becoming more serious and, in 1971, that by small birds was greater than ever before.

Measures to decrease house-sparrow numbers around farm buildings are worth while (and not only for the sake of the sugar beet crop), but continuous action is necessary as the birds soon re-infest a cleared area. Methods commonly used are cage trapping, netting, shooting and the destruction of nests. It is also possible for *bona fide* pest control operators to obtain a licence to use stupefying baits to control house-sparrows.

The other birds that feed on seedling beet do so early in the morning; they may not be seen and their depredations are not at first noticed. Although it is difficult to deter them, especially in large fields, it is worth trying with skylarks since, if successful, they will adopt a new territory and probably not return that season.

Bird Pests: Control, Deterrence, Repellence

Measures specific to the individual species have already been given, but measures common to all are considered in this section. Birds are quick, wary and mobile and are able to learn quickly; the pest species are therefore difficult to control, or to deter or repel for any length of time. Direct action against birds may only be taken within the limits allowed by various Acts of Parliament (see page 91). Broadly, these forbid indiscriminate killing by gun or trap, and prohibit the use of poison or stupefying baits except under licence because they might destroy many kinds of wild life and might be a danger to stock. The use of cage traps is permitted and these should be visited at least once daily for the removal and humane destruction of harmful birds; protected species should be released unharmed. In addition, game birds are subject to the Game Acts and the landlord may impose special conditions of his own.

It is well known that flocking birds can be difficult to deter permanently from visiting a field they favour. If there is no alternative food available locally, the deterrent may be ineffective. Scarecrows, carbide guns, cartridge ropes, and similar manufactured or home-made devices that deter visually or audibly, are usually of temporary value; for best effect they must be moved frequently and their pattern of action altered. By these means it may be possible to protect susceptible seedlings, but by-laws may prohibit explosions during the hours of darkness, and in daytime they are unpopular near houses. Model hawks suspended as high as possible or hydrogen-filled neoprene balloons flown at 50–100 ft have been used with some success to protect cereal trial plots, but such devices are not practical for large fields.

Many chemical repellents have been tested. Anthraquinone seed-dressing can discourage birds from eating the seed but not the shoot of the seedling; it and thiram are of little value when sprayed on sugar beet foliage, except perhaps in the seedling stage and then only very temporarily. Spray materials used for insect control probably repel birds for a day or two; incorporation of repellent materials in the seed pellet is being tested.

Bird problems in sugar beet, and indeed in field and garden crops generally, are largely unsolved.

## RABBIT*

The adverse effects of grazing by rabbits on cereals and many other crops are well known. On sugar beet they feed on the leaves of seedlings but prefer the crowns of larger plants, severing them and partially eating the leaf petioles. They are untidy feeders and usually scatter the severed leaves (c.f. hare damage, page 97). Sometimes they also gnaw into and round the crown, causing damage similar to that caused by pheasants (see page 94). Tell-tale faecal pellets, scratch marks and shallow holes dug into the ground indicate the presence of rabbits; damage is nearly always worse near burrows or shelter.

In 1952 myxoma virus was introduced unofficially into France and it spread rapidly and uncontrollably. The virus gained entry into Britain and the first recorded outbreaks of myxomatosis were confirmed in Kent and East Sussex in October, 1953. Although the outbreak areas were wired off and the rabbits within them killed, it soon became clear that the disease could not be contained. The main agent carrying the virus between rabbits in this country is the rabbit flea. In 1954 and 1955 the virus spread throughout Britain. Very few rabbits that caught the disease survived, but pockets of rabbits that escaped infection continued to live and breed. Secondary outbreaks followed in 1956 and 1957, and rabbit populations throughout Britain were very small and damage to crops was negligible. Outbreaks have continued but less virulent strains of the virus have appeared and some rabbits are resistant. Here and there rabbits have begun to increase again and continue to live above ground where there is sufficient cover.

At present rabbit populations are still relatively small and it is important to prevent them increasing. In 1958 local Rabbit Clearance Societies were started to co-ordinate control measures. Gassing is permissible and effective.

---

* See also Advisory Leaflet 534, available from the Ministry (p. 108).

The best technique is first to block all holes with earth; within a few days occupied burrows are opened by the rabbits and calcium cyanide powder is then spooned 6 in. down the opened holes, or powder is forced down the holes with a pump, and the entrances again blocked. Under the action of moisture, hydrogen cyanide is released and kills the rabbits. If any holes are re-opened the process is repeated. Methodical trapping or ferreting also helps control, but the Pests Act, 1954, stipulates that only humane traps may be used and these must be placed in burrows. Rabbits living above ground may be snared or shot.

None of these measures exterminates rabbits permanently from an area, for immigrants soon begin recolonization. Therefore, to keep rabbit numbers down all available measures should be taken throughout the year, and over-grown hedgerows, tracts of scrubland and wasteland must either be cleared or fenced off.

## HARE

The hare is common in all beet-growing counties, especially where fields are large and dwellings few. It is a solitary animal except during the mating season in February and March. It feeds on beet in much the same way as the rabbit, grazing on seedlings, but prefers the petioles and heart leaves once plants are in the rough-leaf stage. The hare may eat off the outer leaves to get at the heart leaves (Plate XII) and sometimes feeds on the flesh of the crown. Damage can be distinguished from that caused by the rabbit because it is distributed uniformly over the whole field and not concentrated near shelter. The hare does not scatter the leaves and rarely scratches or digs in the soil. The faecal pellets of the hare are also distinctive, being larger and greyer than those of the rabbit.

Numbers of hares fluctuate from year to year and in areas where hares are likely to be troublesome numbers are best kept down by well-organized shoots in February and March. Systematic use of snares in fences and hedges can be very effective. These should be constructed of eight-strand brass wire and set higher on the run than for rabbits.

## COYPU*

The coypu or nutria is a large brown rodent with orange incisor teeth. When fully grown it may be nearly two feet long, excluding the tail, and weigh up to 20 lb. It spends much time in the water, where it swims actively using its webbed hind feet. Large burrows are made in dyke and river banks, sometimes undermining bridges and roadsides. Coypus are mainly active at night, raiding fields of cereals, kale and sugar beet and causing much damage when they are numerous. Sometimes they also gnaw tree roots, but their food is mainly water plants and occasionally molluscs such as the fresh water mussel. Because coypus are rarely seen by day, their presence is usually only indicated by fresh droppings, recently used runs, grazed vegetation or damaged crops. Typically, young beet plants are bitten off, but the exposed parts of mature beets may be chewed, with or without the crown being removed.

---

* See also Advisory Leaflet 479, available from the Ministry (p. 108).

The coypu was introduced into Britain about 1929 and farmed for its fur. Inevitably some escaped and by 1944 coypus were firmly established in East Anglia, especially in the favourable areas of the Broads, and from there they spread along many East Anglian rivers. There was a danger of their gaining a foothold in the Fens so that in 1962 an organized campaign was launched against the coypu. Stretching in arcs from the Wash to the coast of Suffolk, successive four-mile belts were cleared eastwards. The operation was planned from a coypu control centre in Norwich and carried out by Ministry of Agriculture trappers with the help of Rabbit Clearance Societies and River Boards. The object was to rid East Anglian rivers of coypus and to confine them to the Broads, where the terrain is greatly in their favour and where it may not be possible to exterminate them. The hard weather in January and February, 1963, killed many coypus but hindered the campaign by making survivors disperse.

It is probable that coypus are not 'small ground vermin' within the meaning of the Protection of Animals Acts, 1911 and 1927, and therefore poisoning coypus, though feasible, is at present illegal. In the U.S.S.R. coypus have been lured to rafts by baiting and then trapped. In Britain coypus may be trapped in cage traps placed on or near their runs and emergence sites along rivers, dykes, broads and similar situations. They may also be shot, but their nocturnal habit limits this method. The trapping campaign of 1962–65 was very effective and the subsequent joint efforts in East Anglia have so far managed to contain the survivors.

## FIELD MICE

In 1971, field mice (*Apodemus* spp.) caused damage in many beet-growing areas by digging out and eating not yet germinated seed; such damage had occasionally been noted in previous years but only on an insignificant scale. In 1971 loss of seeds was sometimes confined to the field margins but often occurred in the middle of fields; mouse holes could be found in some fields. Damage was severe where seed was spaced 5 in. or more apart and long lengths of row were destroyed, even leading to the need to re-drill. The damage is characterized by the accuracy with which the mice locate the buried seed in the row, the neat holes dug down to the seed, the breaking of the seed pellet where present, and the removal of the seed cap leaving the fruit body with empty loculus. Field mice can readily be caught in the field with baited traps, and masticated seed embryos can be recognized in their stomachs.

The widespread damage in 1971 may have been due to a combination of unusual circumstances, for instance more mice than usual having over-wintered and the beet seed germinating slowly due to cold, dry conditions.

Control measures—trapping, baiting, gassing—need to be applied as soon as any damage is noted.

## OTHER MAMMALS

*Moles* are often troublesome in beet fields in April and May, especially in soils where earthworms, wireworms and other soil creatures are numerous. The moles, in their search for food, tunnel long lengths of row just below the soil surface; they destroy more seedlings than do the various soil creatures they are seeking, except in the case of chafer grubs when moles can be beneficial. If mole-infested permanent pasture is ploughed up, mole damage is almost inevitable in the crop that follows. If at all possible the pasture should be cleared of moles before ploughing. Alternatively, prompt trapping or poisoning as soon as the runs are noted in the crops can prevent many seedlings being lost (see Advisory Leaflet 318*).

*Rats.* In summer and autumn, rats (see Advisory Leaflet 516*) chew the petioles and crowns of beet near hedgerows and farm buildings; in autumn and winter they feed on clamped beet. Rats should be controlled whenever possible by bait-poisoning, gassing, trapping or shooting. Gassing, by using calcium cyanide powder as for rabbits (see page 97), is the most effective method of control when rats are in burrows.

*Deer* also feed on beet and can be very troublesome locally.

## FURTHER READING ON BIRDS AND MAMMALS

ANON. (1962 and 1965). *Infestation control.* Minist. Agric. Fish. Fd, Lond., H.M.S.O.

DAVIS, R. A. (1970). Control of rats and mice. *Bull. Minist. Agric. Fish. Fd, Lond.* No. 181. H.M.S.O.

DAVIS, R. A. (1963). Feral coypus in Britain. *Ann. appl. Biol.* **51**, 345–8.

LACK, D. (1954). *The natural regulation of animal numbers.* Oxford, Clarendon Press.

LANCUM, F. H. (1948). Wild birds and the land. *Bull. Minist. Agric. Fish. Fd, Lond.* No. 140. H.M.S.O.

MURTON, R. K. (1960). Population dynamics of the wood-pigeon (*Columba p. palumbus*) and methods of control. *Ann. appl. Biol.* **48**, 419–22.

MURTON, R. K. (1962). Narcotics *v.* wood-pigeons. *Agriculture, Lond.* **69**, 336–9.

MURTON, R. K. (1965). *The wood-pigeon.* New Naturalist Special Volume M. 20. London, Collins.

MURTON, R. K. (1971). *Man and birds.* New Naturalist Series, No. 51. London, Collins.

NICHOLSON, E. M. (1951). *Birds and men.* New Naturalist Series, No. 17. London, Collins.

NORRIS, J. D. (1967). A campaign against feral coypus (*Myocastor coypus* Molina) in Great Britain. *J. appl. Ecol.* **4**, 191–9.

SOUTHERN, H. N. (Ed.) (1964). *The handbook of British mammals.* Oxford, Blackwell Scientific Publications.

THOMPSON, H. V. (1958). Rabbit control in Australia and New Zealand: 1. *Agriculture, Lond.* **65**, 388–92.

THOMPSON, H. V. (1960). Economic ornithology. *Ann. appl. Biol.* **48**, 405–8.

THOMPSON, H. V. (1961). The history and present situation of myxomatosis. *Penguin Books Science Survey* No. 2, 102–11.

THOMPSON, H. V. (1963). The limitations of control measures. *Ann. appl. Biol.* **51**, 326–9.

THOMPSON, H. V. and WORDEN, A. N. (1956). *The rabbit.* New Naturalist Special Volume M. 13. London, Collins.

* Available from the Ministry (p. 108).

H

# Foreign Beet Pests

BRIEF mention must be made of insects injurious to sugar beet abroad that are found in England but are not known to be harmful here. Amongst these are the beet moth (*Scrobipalpa ocellatella* (Boyd)), which has greenish or reddish caterpillars that feed in the heart of the sugar beet, tunnelling into the bases of the leaf petioles and sometimes causing serious losses in southern Germany and in Greece and Italy. The beet webworm (*Margaritia sticticalis* (L.)) defoliates sugar beet in parts of Europe and in the United States of America. The caterpillars of *Hymenia recurvalis* F. attack beet in Europe; the adult was recently collected near the south coast of England. The English summer is probably not warm or dry enough to permit the development of damaging populations of these three insects.

One further species of insect is considered below.

## BEET LEAF BUG

The beet leaf bug (*Piesma quadratum* (Fieb.)) (Plate II) sometimes damages beet in W. Germany and parts of eastern Europe by transmitting the virus disease known as beet leaf curl or crinkle (Krauselkrankheit). The insect is found also in England, chiefly near the coast. It feeds, by piercing and sucking as do aphids, on wild plants such as sea purslane (*Halimione portulacoides* (L.) Aell.), shore orache (*Atriplex iittoralis* L.) and wild beet. The bugs hibernate as adults around the margins of beet fields, and in April and May migrate to the beet fields where they feed on the plants and related weeds. Egg laying begins in May and continues until August; there is a partial second generation. Although the eggs are small, they may be found readily on the veins on the underside of the leaves. They are elongate and, when first laid, are cream but later turn yellow and finally brown. The young bugs are yellow but become greener after moulting twice and acquiring wing buds. Adults are grey or brownish with characteristic lace-like markings on the upper surface of the body.

Beet leaf curl is unknown in Britain. It was first observed in Silesia in 1903 and later spread to parts of north and east Germany and Poland. Each year infected beets were found further west in Germany, since the infective bugs are blown by the strong east winds in the spring. The adults overwinter with the virus in their salivary glands and transmit the disease as soon as they feed on host plants after hibernation. The worst injury is near hedgerows and other places where they shelter during the winter. Attacked seedlings may be killed; older plants first show symptoms of vein clearing, which is best seen by holding leaves up to the light. Later the veins swell and the leaves crinkle. The leaves also curl inwards and, as the plant grows, a definite heart is developed reminiscent of a cabbage or lettuce. The yield is greatly decreased.

The closely related bug *Piesma maculatum* (Lap.), common on beet crops in the Breckland and elsewhere in England, does not transmit beet leaf curl virus.

## FURTHER READING ON FOREIGN BEET PESTS

ANON. (1968). *Les principaux parasites et maladies de la betterave sucrière*. Paris, Institut Technique Français de la Betterave Industrielle.

BONNEMAISON, L. (1948). Les parasites animaux de la betterave. *Cah. Inst. tech. Betterave* No. 4.

DUNNING, R. A. (1972). Sugar beet pest and disease incidence and damage, and pesticide usage, in Europe: report of an I.I.R.B. enquiry. *J. int. Inst. Sug. Beet Res.* (In the press).

JOHNSON, R. T., ALEXANDER, J. T., RUSH, G. E. and HAWKES, G. R. (Eds.) (1971). *Advances in sugarbeet production: principles and practices*. Ames, U.S.A., Iowa State University Press.

KERSHAW, W. J. S. (1964). *Piesma quadratum* (Fieb.) in East Anglia. *J. agric. Sci., Camb.* **63,** 393–5.

PETHERBRIDGE, F. R. and STIRRUP, H. H. (1934). Investigations in 1934 of sugar beet pests and diseases: report on a Continental tour to Informal Committee on Sugar Beet Research and Education. *Minist. Agric. Fish. Comm. Pap.* No. 31.

# Other Troubles Confused with Pest Damage

BULLETIN 142 deals with diseases of sugar beet. Various causes of damage sometimes confused with pest damage are considered below.

## STRANGLES*

From late May to July plants sometimes break off at, or just below, ground level as though eaten through (Plate VIII). Insects, especially ground beetles, may be found feeding on sap exuding from the stump, or mould organisms may develop there, but both are secondary. The trouble is called strangles and is caused by a constriction which develops at about soil level (Fig. 4(b)) and becomes more pronounced as the root grows and swells.

The primary cause is drying of the newly exposed upper root of the seedling immediately after singling; this is most likely to occur when much soil is removed from round the plants as they are singled in the cotyledon and two rough-leaf stage. A 'stringy root' develops and, as the plant contracts and pulls itself back into the soil, this remains constricted while thickening occurs above and below. The drying of young, exposed seedlings by wind induces the trouble. Wind also causes strangles in larger plants by flexing the root at soil level two or three weeks after singling.

The effect of strangles may not be seen until plants topple over in the wind or during steerage hoeing. Losses start about two weeks after singling and continue into July, but fortunately are scattered throughout the crop and do not affect yield greatly when the plant population is adequate. As many as ten per cent of plants may be lost but this is unusual.

Soil-incorporated herbicides sometimes cause severe twisting which 'strangles' the root deep in the soil, but the affected plant rarely dies (Plate VIII).

Strangles must not be confused with damage caused by hoeing or by pheasants. With strangles, the skin of the root is intact (Fig. 4(b)). Pests, such as pygmy beetle, that damage seedling roots do not induce strangles. During secondary thickening of the root the skin of the hypocotyl splits naturally; this is not connected with strangles nor should it be confused with injury by insects or chemicals.

FURTHER READING

BOYD, A. E. W. (1966). Sugar beet strangles. *Tech. Bull. Edinb. Sch. Agric.* No. 26.

BOYD, A. E. W., ERSKINE, D. S. C., BYFORD, W. J. and WEBB, D. J. (1970). A herbicide-induced abnormality in sugar beet. *Pl. Path.* **19**, 163–4.

## HERBICIDE DAMAGE

The use of herbicides on beet, and indeed on all agricultural crops, is increasing. Great care must be taken in applying them because unwanted

---

* See also Advisory Leaflet 547, available from the Ministry (p. 108).

side effects may arise. Herbicide damage to beet may be caused in three ways:

(1) *Residues from applications to previous crops.* So far as is known, damage from residues is caused in beet only where far too much herbicide has been applied to the previous crop, usually by spillage or where a sprayer has stopped for a time with the spray turned on. The pattern of symptoms in the beet crop often gives a clue to the cause. TBA (2, 3, 6-trichlorobenzoic acid) with MCPA, a post-emergence cereal spray, causes leaf thickening and edge crinkling in beet and potatoes the following year, but only at some ten or more times the normal dose. Similarly, excessive amounts of TCA (trichloro-acetic acid, usually as its sodium salt) applied to the preceding crop, or before sowing, can constrict the beet at soil level and cause warty growth similar to crown gall above the constriction (see Bulletin 142*). Excessive applications of simazine to beans may persist and affect a following beet crop.

(2) *Applications to beet of excessive doses or of inadequately tried chemicals.* Many materials are now used as pre- or post-emergence herbicides on the beet crop and must be applied correctly with great care. Faulty spraying methods are probably the main cause of spray damage. TCA, propham and di-allate are applied before sowing and worked into the seedbed. Under some conditions the first two may cause stunting, and all damage the crop when rates are excessive. Even at normal rates of application di-allate may lead to distortion and fusing of leaves in May and June, but this probably does not harm the beet. Soil-acting herbicides such as endothal, propham, diuron, cycluron (OMU), chlorbufam (BiPC), fenuron, or mixtures of these, and also pyrazon and lenacil are applied after sowing and before emergence. Even normal doses may stunt or retard early growth and, under exceptional soil conditions, may decrease germination, or scorch or kill seedlings after emergence. When these herbicides have been used, pest damage may be accentuated by slow growth and mistakenly suspected as the primary cause; the reverse may, of course, also apply and herbicides should not be blamed unnecessarily. Contact, pre-emergence herbicides such as diquat and para-quat leave virtually no residue in the soil and can be applied immediately before seedling emergence. Other materials such as PCP, dimexan and a dimexan/cycluron/chlorbufam mixture must be applied at least three days before emergence. If applied later, or excessively, abnormal growth may result. The use of a post-emergence herbicide, such as phenmedipham, avoids the risk of the plants absorbing too much herbicide from the soil, but the spray may damage the beet if the leaves are already affected by nutrient deficiency or pest attack. Where more than one herbicide is applied, advice should be sought.

**Sprayers used to apply herbicides should not be used to apply insecticides to beet without the most thorough washing.**

(3) *Drift from applications to adjoining crops.* Spray mist is likely to drift during application to any crop and great care must always be taken to prevent it. The amount of drift depends on weather, but the result of drift depends on the susceptibility of the crops downwind. Sugar beet, especially when growing

---

* Bulletin 142: *Sugar Beet Diseases.* Obtainable from Her Majesty's Stationery Office, or through any bookseller, price 37½p by post 43p.

rapidly, is easily upset by very low concentrations of growth-regulator herbicides such as MCPA and 2,4-D (Fig. 4(c)) and may be upset by many other herbicides. More difficult to avoid than the drift of spray mist is the blowing of dried herbicide deposits, usually from cereal leaves. Also, damaging amounts of vapour may drift with light breezes on hot days, but fortunately this is most unusual under British conditions.

Despite the dramatic effects of some of the growth-regulator herbicides on beet foliage, yield is likely to be decreased only when growth is retarded and not when the plants produce abnormal leaves for a short time.

## WEATHER INJURY

Wind-blown soil particles abrade seedlings and, especially if sandy, can kill them. On soils liable to blow, whole areas of the field may be lost or seedlings buried by drifting soil. Wind and wind-blown particles scar the leaf surface and tear and blacken leaf edges; the only pest causing somewhat similar damage is thrips (see page 64).

Frost injury to young seedlings may cause thickening and curling of the cotyledons. Similar symptoms can be caused by aphids feeding on young seedlings, especially by one or two species of green aphid other than *Myzus persicae*.

Hailstorms in July and August tear the leaves and, if heavy, may defoliate the plants. Some caterpillars cause similar damage, especially those of the silver Y moth (see page 44), but they or their pupae can readily be found in the crop.

## DOWNY MILDEW

This disease yellows the outer leaves so that they are virtually indistinguishable from those yellowed by virus (see page 55). Downy mildew can be confirmed by examination of the heart leaves, which are thickened, stunted, distorted and covered with a grey fungal mycelium when the infection is active. When an active infection is lacking, there are usually signs of shrivelled heart leaves or of a dead growing point.

## BLIND SEEDLINGS

When the primary growing point is killed in the cotyledon or two rough-leaf stage, a blind seedling results. Subsequently the cotyledons and first rough leaves expand and thicken considerably and, at the same time, axillary growing points develop, leading to a multi-crown plant. The primary growing point is rarely killed in seedlings at or beyond the four rough-leaf stage, except by boron deficiency in July or later (see Bulletin 142*).

There are several causes of seedling blindness, the commonest being capsids (see page 63) or stem eelworm (see page 67). Other causes are herbicide toxicity, bird or mammal damage, or mechanical injury.

---

\* See footnote, page 103.

# Beneficial and Other Insects not Attacking Beet

MANY insects other than pests occur in beet crops throughout the year and are either beneficial or harmless.

Beneficial insects are those that benefit the crop, usually because they eat pest insects. They may be predators already in the crop, such as ground or rove beetles and their larvae, or they may be migrant predators that move into the crop in search of food for their larvae only (e.g., hover flies) or for themselves and their larvae (e.g., lacewings and ladybirds). The last three feed almost exclusively on aphids, laying their eggs on the leaves when aphids are present or before they arrive. Adult ladybirds often help to check aphid infestations in June, and larvae of all three insects help to destroy aphid populations in the crop in July (see page 47). Active, white, legless, wiry larvae known as 'white wireworms' may be found when soil is disturbed; they are the larvae of a fly (family *Therevidae*) and do not feed on plants but on other insects in the soil. Another type of beneficial insect is one that parasitizes its host; all insects have parasites. When pests are numerous, predators and parasites multiply rapidly, causing a sharp decline in pest numbers. Parasites of beet leaf miner have been studied in detail. The egg stage is parasitized by at least one species of insect and the larvae and pupae by at least eleven others. Up to 90 per cent of pupae collected from the soil during the winters of 1949, 1950 and 1951 contained parasites or, occasionally, predators and not beet flies. This degree of parasitism or predation helps greatly to control pest numbers, although unfortunately the beneficial species do not always become numerous enough to exert effective control until serious harm has been done to a crop.

A different kind of beneficial insect is the polygonum leaf beetle (*Gastrophysa polygoni* (L.)) (Plate II), commonly found in beet fields from April to June. This conspicuous globular insect, about ¼ in. long, with bright greenish-blue wing cases, has been suspected of eating beet, but it feeds on various weeds, of which knotgrass (*Polygonum aviculare* agg.) and black bindweed (*Polygonum convolvulus* L.) are the chief; it is therefore beneficial.

The remaining group of insects found in beet fields, the harmless or neutral ones as far as beet is concerned, are rarely very numerous. Some are common pests of the previous crop, while some may have flown or been blown there during dispersal flights (e.g., the pollen beetle, commonly confused with flea beetle); they do not stay long as there is no food for them. Others may be insects from woodland or hedgerows and are there by chance temporarily, while the remainder or their larvae are probably feeding on decaying organic matter in the soil.

FURTHER READING

MORETON, B. D. (1969). Beneficial insects and mites. *Bull. Minist. Agric. Fish. Fd, Lond.* No. 20. H.M.S.O.

---

Pest damage to beet is often characteristic and readily identifiable, but it is usually valuable to be able to confirm diagnosis by finding the pest in the crop and sometimes this is essential (e.g., eelworms, symphylids). When the cause of injury to plants is uncertain, the occasional insect that can be found should not be blamed. Any insects or other creatures found in the soil or crop must be actually feeding or able to feed on the crop and must be sufficiently numerous to do harm. All likely pests of sugar beet and some beneficial insects are described in this Bulletin and many are illustrated. If injury is occurring and pests cannot be identified, a specialist should be asked to identify those creatures that can be found (see page 11).

# Metric Equivalents of Non-metric Units

| Non-metric Unit | | | | Metric Equivalent |
|---|---|---|---|---|
| in. | .. | .. | .. | .. 25·4000 mm |
| ft | .. | .. | .. | .. 0·3048 m |
| yd | .. | .. | .. | .. 0·9144 m |
| mile | .. | .. | .. | .. 1·6093 km |
| sq. yd | .. | .. | .. | .. 0·8361 sq. m |
| acre | .. | .. | .. | .. 0·4047 ha |
| sq. mile | .. | .. | .. | .. 2·5899 km² |
| fl. oz | .. | .. | .. | .. 28·4125 ml |
| pint | .. | .. | .. | .. 0·5683 l. |
| gal | .. | .. | .. | .. 4·5461 l. |
| oz | .. | .. | .. | .. 28·3500 g |
| lb | .. | .. | .. | .. 0·4536 kg |
| cwt | .. | .. | .. | .. 50·8032 kg |
| ton | .. | .. | .. | .. 1·0160 t |

## Conversion Factors

plants/acre $\times$ 2·47 = plants/ha
fl. oz/acre $\times$ 70·20 = ml/ha
gal/acre $\times$ 11·23 = l./ha
oz/acre $\times$ 70·05 = g/ha
lb/acre $\times$ 1·12 = kg/ha

# Precautions

WHENEVER pesticides are used, read and follow carefully the instructions on the label. Users should also consult the 'Recommendations for the Safe Use of Chemical Compounds Used in Agriculture and Food Storage' published by this Ministry. Use the pesticides mentioned in this Bulletin with care, particularly:

| | | |
|---|---|---|
| calcium cyanide | endothal* | oxydemeton-methyl* |
| demephion* | ethoate-methyl | paraquat |
| demeton-S-methyl* | fenitrothion | Paris green |
| di-allate | formothion | pentachlorophenol (PCP) |
| dichloropropane— | lead arsenate | phenmedipham |
| dichloropropene mixture | lenacil | phorate* |
| dichloropropene | malathion | phosphamidon* |
| dieldrin | menazon | thiometon* |
| dimethoate | methiocarb | trichloroacetic acid (TCA) |
| diquat | nicotine* | trichlorphon |
| disulfoton* | | |

and wash off any concentrate that falls on the skin. Store new and part-used containers in a safe place under lock and key. Empty metal or plastic containers should be thoroughly rinsed out before the spray tank is completely full and the rinsings poured into the spray preparation. Cardboard cartons should be emptied completely and burned. If the container will not burn it should be flattened and buried. Do not contaminate ponds, waterways or ditches with chemical or used containers.

---

* There are statutory obligations affecting employers and workers who use certain poisonous substances, including demephion, demeton-S-methyl, disulfoton, endothal, nicotine, oxydemeton-methyl, phorate, phosphamidon and thiometon. Users of these chemicals are strongly advised to read the Ministry's leaflet APS/1, 'The Safe Use of Posionous Chemicals on the Farm,' obtainable free from the Ministry at the address given below. Users of hydrogen cyanide and methyl bromide should also comply with the Home Office regulations governing the use of these chemicals for fumigation. The Home Office regulations are obtainable, price 2½p (by post 4p), from Her Majesty's Stationery Office at any of the addresses on the back cover.

## AGRICULTURAL CHEMICALS APPROVAL SCHEME

Proprietary products based on chemicals used for pest, disease and weed control can be officially approved under the Agricultural Chemicals Approval Scheme. It is strongly recommended that approved products should be used. Approval is indicated on the containers by the mark shown here. A List of Approved Products is published in February of each year and is obtainable free of charge on application to the address below or any Regional or Divisional Office.

AGRICULTURAL CHEMICALS
APPROVAL SCHEME

Advisory Leaflets are obtainable free from the Ministry of Agriculture, Fisheries and Food (Publications), Tolcarne Drive, Pinner, Middlesex HA5 2DT.

# Index

Printed in England by Her Majesty's Stationery Office at HMSO Press, Harrow

165255 Dd 502644 K 24 10/72

MINISTRY OF
AGRICULTURE, FISHERIES AND FOOD

# Publications on
# Pests and Diseases

| | | |
|---|---|---|
| Common Names of British Insects and Other Pests. (Tech. Bulletin 6.) | 50p | (54½p) |
| Narcissus Pests. (Bulletin 51) | 65p | (69½p) |
| Diseases of Bees. (Bulletin 100) | 35p | (37½p) |
| Diseases of Bulbs. (Bulletin 117) | 25p | (30½p) |
| Diseases of Vegetables. (Bulletin 123) | 62½p | (69p) |
| Cereal Pests. (Bulletin 186) | 60p | (65½p) |

*Price in brackets includes postage*

Published by
## HER MAJESTY'S STATIONERY OFFICE

and obtainable from the Government Bookshops in London (post orders to P.O. Box 569, SE1 9NH), Edinburgh, Cardiff, Belfast, Manchester, Birmingham and Bristol, or through booksellers